Atomic Spectroscopy

Atomic Spectroscopy

JAMES W. ROBINSON
Department of Chemistry
University of Louisiana
Baton Rouge, Louisiana

MARCEL DEKKER, INC.   New York and Basel

Library of Congress Cataloging-in-Publication Data

Robinson, James W.
    Atomic Spectroscopy / James W. Robinson.
      p.   cm.
    Includes bibliographical references and index.
    ISBN 0-8247-8311-5 (alk. paper) :
    1. Atomic spectroscopy. I. Title.
QD96.A8R63   1990
543'.0858--dc20   90-3452
                                 CIP

This book is printed on acid-free paper.

MARCEL DEKKER, INC.
270 Madison Avenue, New York, New York 10016

Current printing (last digit):
10 9 8 7 6 5 4 3 2 1

PRINTED IN THE UNITED STATES OF AMERICA

Preface

This book covers atomic spectroscopy in the electronic excitation region of the electromagnetic spectrum. Specifically, the fields covered are flame photometry, atomic absorption, atomic fluorescence, emission spectrography, plasma emission, and the allied tandem instrument interfaced plasma-mass spectrometry. Each topic is addressed individually with some acknowledgment of the historical development of the physical process and its conversion to an analytical procedure.

The first chapter discusses atomic energy transitions, their probability, and the population of the various states as described by the Boltzmann distribution. It also describes free atom formation, optical systems, error analysis, Beer's law, signal-to-noise ratio, and sampling techniques. This chapter provides a springboard for the other chapters.

Each subsequent chapter describes a physical phenomenon and its development into the pertinent analytical technique. The respective equipment is described together with sources of analytical error and methods of overcoming them. In each chapter the virtues and shortcomings of the procedures are discussed. The different techniques are finally compared with one another with respect to the sensitivity limits, the analytical range, and simultaneous, sequential, or one-at-a-time analysis capability.

The book should be most useful to the practicing analytical chemist in a routine industrial or research environment. It would also be useful in a university setting as a text for a special topics course.

James W. Robinson

Contents

Preface   *iii*

1   Introduction   **1**
     History of Atomic Spectoscopy   1
     Atomic Spectroscopy as a Tool for Elemental Analysis   3
     Comparison of Absorption and Emission   18
     The Beer-Lambert Law   24
     Summary of the Comparison of Absorption and Emission   25
          Techniques
     Optics in Spectrophotometry   26
     Reliability of Results   30
     Signal and Noise (Sensitivity)   47
     Sampling   50

2   Atomic Absorption Spectroscopy   **55**
     General Principles   56
     Equipment Used for Atomic Absorption   56
     Analytical Parameters   87
     Recommended Procedures for Quantitative Analysis   141
     Electrothermal Atomization   146

Contents

3   Atomic Fluorescence   **167**
Introduction   167
Mathematical Relationships   168
Advantages of Atomic Fluorescence   171
Limitations of Atomic Fluorescence   172
Atomic Fluorescence as an Analytical Tool   172

4   Flame Photometry   **189**
Origin of Spectra   191
Equipment   195
Flame Emission   204
Analytical Applications   207
Conclusions   213

5   Emission Spectroscopy   **215**
History   215
Relationship to Flame Photometry   217
Equipment   218
Analytical Applications   230

6   Plasma Emission   **241**
Introduction   241
Equipment   245
Analytical Applications   264
Interfaced ICP—Mass Spectrometer   281
Conclusions   288

Index   *295*

Atomic Spectroscopy

1

Introduction

A. HISTORY OF ATOMIC SPECTROSCOPY

The history of atomic absorption and the history of atomic emission are intimately interwoven. The story begins with the description of the visible spectrum by Isaac Newton (1). This early work generated studies on the interpretation of the physical properties of atoms.

For about 100 years the studies were entirely focused on the emission of radiation from atoms. But in 1802 Wollaston (2) reported the presence of dark bands in the radiation from the sun. Later Fraunhofer in 1814 and Brooster in 1823 measured the position of several hundred of these and designated the origin of many of them.

At first it was thought that the absence of these atomic lines in the sun's spectrum indicated lack of this radiation and therefore of the relevant elements on the sun's surface. Foucault in 1849 passed sunlight through a flame containing sodium vapor, expecting the dark sodium (Fraunhofer) lines to become more intense, indicating atomic absorption. Kirchoff in 1859 deduced that the dark lines in the sun's spectra were caused by absorption by atomic vapors rather than lack of atomic emission because the elements were absent. He termed these dark lines "reversed spectra."

Forty years later Kirchoff and Bunsen showed that the wavelengths of one of these dark lines corresponded exactly with the yellow emission line obtained from excited sodium vapor.

As a result of extensive studies, Bunsen and Kirchoff posited the fundamental principles:

1. Every element when sufficiently excited in the gaseous state yields its own characteristic spectrum consequently.
2. The presence of the vapor of an element can be inferred with certainty by spectrum analysis when its characteristic lines are present.

Based on this work Kirchoff further developed the important relationship that any species that can be excited to emit radiation at a particular wavelength can absorb at that wavelength. This pronouncement showed the fundamental relationship between emission and absorption spectra. For many years the study was confined to the interpretation of the spectra in solar and interstellar atmospheres and the elucidation of which elements were present in those atmospheres. By 1930 the fields of emission spectroscopy including arc and spark emission were well established but were used primarily for qualitative analysis. Practical difficulties were encountered with absorption processes including the use of a photographic plate and the necessity of using high resolution systems. However, this was changed in 1953 by Alan Walsh (3) and confirmed by Alkamde and Melatz in 1955.

Walsh's contributions to the field of atomic absorption were significant. He used the hollow cathode as a light source, thus greatly reducing the resolution required for successful analyses. Second, photomultipliers were now available, and their use eliminated the problems associated with measuring small amounts of absorption with a photographic plate that necessitated the detection and measurement of minor changes in darkness of this line (comparing a dark line with a slightly less dark line). Third, he introduced modulation into the system, which permitted the detector to distinguish between absorption by the atoms and emission by the atoms at the same wavelength. Fourth, he utilized a flame as an atomization system. These contributions were all major and vital to the development of atomic absorption as we know it today as an analytical tool.

At that time the author was most fortunate to learn an eternal truth that governs the advancement of analytical instrumentation. In 1956 there was a conversation at my home between Sir Alan Walsh and A. J. P. Martin, N.L., of chromatography fame. At that time the field of atomic absorption had not been accepted by the instrument manufacturers and was fast going nowhere. Dr. Martin related his experiences in promoting chromatography. He said that in earlier work he had described the theory and potential of chromatography but the scientific community completely ignored it, as they had Tswett before him. Later he published an article describing the separation of the amino acids using this technique. This was a clear demonstration of a very valuable application of the technique. Immediate worldwide attention resulted, together with the award of a Nobel Prize. The important message was, "Theory is fine, but solving an important problem is what gets attention." Ideally commercialization follows, otherwise the technique resides in the archives and slowly rots away.

Dr. Walsh was not an analytical chemist, and although he had shown the potential of a technique, he had not unequivocally demonstrated its analytical usefulness. After subsequent discussions with several instrument companies, the instrument manufacturers remained unimpressed. However, on visiting the author's laboratories at Esso Research in 1957, Perkin Elmer Corporation, represented by John Atwood, found that the technique was already being used for routine analysis. This got their attention. Immediately blueprints were drawn up and within a fairly short time the PE models 290 and 303 came on the market and the rest is history.

The moral is that someone has to demonstrate practical analytical use to prove to the practicing bench chemists that this "exciting new technique" with "lots of potential" really works and provides sought-after data. Only then will the manufacturers develop and sell such equipment. Until that happens the technique remains among the list of "potentially useful, physical phenomena" and dies on the vine within a couple of years.

B. ATOMIC SPECTROSCOPY AS A TOOL FOR ELEMENTAL ANALYSIS

The determination of the concentration of a particular element in a sample irrespective of the chemical form of that element is part of the broad analytical field of "elemental analysis." For many years the procedures used in analytical chemistry were overwhelmingly dominated by reliance on the chemical reaction of the elements, and the fields of gravimetric, volumetric, and electrochemistry were widely used in this respect. Usually the methods were time-consuming and subject to numerous errors, many of which could only be eliminated or corrected for by a high degree of skill and tender loving care on the part of the operator—an uncertain quantity at all times. Because of the long lapse time involved and the meticulous attention needed, only the most essential analyses were performed on industrial products by the manufacturing company. Frequently the methods could not be used for product control. Often by the time the analysis was completed, the product was already shipped. Analysts who were really artists in their own right carried out analysis by intuition. Colorimetric procedures were common in which the analyst dissolved the sample and developed a color specific to the component being analyzed—held the solution up to the sky and came to a "go, no-go" decision on the quality of the sample (the "Blue Sky" technique). Such analysts were invaluable to the company, but like many other skills, competant operators often were not replaceable on retirement. With time, less intuitive instrumental techniques were developed which could be successfully operated by much less skilled operators.

In the last five decades there has been a steady drift away from the classical methods of analysis to spectroscopic and other instrumental techniques. For elemental analysis these techniques often depend on the physical rather than chemical properties of the elements. For example, using X-rays many samples can be analyzed with a minimum of preparation. Other processes have been developed which reduce the elements to their atomic form in a highly reproducible manner and determine the concentration of atoms present. Instruments have been developed with sophisticated computer programs which are able to perform with a minimum of input by the analyst. This has led to an explosion in the number of analyses performed on a daily basis, both by industrial manufacturers and other users. Online, continuous analysis is commonplace, and procedures which were once highly complicated have now been programmed to be machine operated.

The good news is that the analytical procedures are computer controlled independent of "operator error" and are highly reproducible. There is a minimum of operator control needed, and the procedures are very rapid, allowing almost real time data to be obtained. The methods work very well for repetitive samples. Such samples must be very similar in all respects to each other.

The bad news is that the computer-controlled analytical procedures do not work well if the sample composition varies significantly. Unfortunately, sample composition often varies somewhat from one sample to the next—that is why we analyze them. This variation may be minor and accurately accommodated by the procedure. But the variation may be major, introducing errors which the computer does not correct; an unskilled operator may not recognize the problem and precautions are not taken to eliminate unsuspecting sources of error.

In the last analysis the computer is always a man-made device which is most successful under very reproducible conditions. The program must be overridden when the conditions vary. To do this the skilled operator must understand the processes involved in the measurements the instruments are taking. Such understanding can mean the difference between generating accurate or inaccurate answers, the importance of which cannot be overemphasized. It becomes obvious if we think of the clinical analyses of a patient's blood, urine, or other body fluids, the results of which determine the medical treatment that must be dispensed to a patient. Inaccurate analyses may result in incorrect medical treatment.

Accurate information is equally important in industry. The presence of certain metals may be vital for the performance of a catalyst for the desired beneficial effect of a pharmaceutical compound. Similarly, some elements are known to be poisons to industrial catalysts or to industrial products, and steps must be taken to eliminate them. It may be equally important that certain elements be absent from foodstuffs. Some elements such as mercury, cadmium, arsenic, etc. are known to be toxic and should not be ingested.

Inaccurate analyses can lead to inaccurate information and incorrect medical or management decisions. The importance of the role of the analyst cannot be overemphasized in medicine or in quality control. To that end it is important to understand the factors that control elemental analysis. Some of these factors will be discussed below.

1. Atomic Theory

The properties of atoms were first comprehensively described by John Dalton (1766–1844). His postulations have been modified over the last century, but the most important components are as follows:

1. All matter is made up of unit particles.
2. There are 103 different kinds of atoms. Each different atom is an element. (We may create more in the future.)
3. The atoms of each element have a fixed assigned atomic number. This number equals the number of protons or electrons in the atomic structure.
4. The atoms of each element have one or more mass values. The number of mass values is small; each is called an isotope.
5. Atoms unite together in many combinations to form compounds. The formation and properties of these compounds make up the chemistry of the element.

a. Components of the Atom

The atom is the smallest unit of a particular element. However, physicists have found that the atoms themselves are composed of smaller particles (subatomic particles).

The atom is composed of a nucleus and a cloud of surrounding electrons. The nucleus contains the bulk of the weight of the atom and is composed of neutrons and positrons. To the first approximation the mass of the nucleus is equal to the sum of the masses of the protons and the neutrons.

The charge of the nucleus is carried by the protons and the charge is nominally equal to the number of protons present in the nucleus. This charge is balanced by the negatively charged electrons. In any particular atom there are an equal number of protons and electrons present, and the total atom is neutral in charge. Each element, therefore, may have a fixed charge on its nucleus (equal to the atomic number). For example, carbon has 6 protons, therefore a positive charge of 6 on its nucleus.

For a particular element the number of protons is fixed. However, the number of neutrons is not fixed, and therefore each element may have a variable number of neutrons present. The mass of the nucleus depends on the sum of the protons and number of neutrons. For example, the mass of carbon is 12, 12 being the weight of 6 protons and 6 neutrons. However, we know from experiment that approximately 1.1% of all carbon atoms contain 6 protons and 7 neutrons with a mass of 13. Further, one in 10^{10} contains 6 protons and 8 neutrons for a mass of 14. Therefore, naturally occurring carbon may have charge equal to 6 and mass equal to 12, 13, or 14 with a relative abundance of 100 to 1.1×10^{-10}. These are written $^{12}C_6$, $^{13}C_6$, $^{14}C_6$ and are isotopes of carbon.

The number of electrons present in an element is always equal to the number of protons present. However, the way in which the electrons are spaced about the nucleus is of great importance since that directly affects the configuration of the free atom.

b. The Weight and Size of the Atom

A cornerstone of atomic theory was provided by Avogadro, who determined that the atomic weight of an element in grams contained 6.02×10^{23} atoms, whatever the element. The relative weights of the atoms of different elements is therefore the ratio of the atomic weights of those elements. For example, 12 g of $^{12}C_6$ contains 6.02×10^{23} atoms of C, 16 g of $^{16}O_8$ contain 6.02×10^{23} atoms of O. Conversely one atom of $^{12}C_6$ weights 12–6.02×10^{23} g or approximately 2×10^{-24} g and one atom of $^{15}O_8$ weighs $16 \div 6.02 \times 10^{23}$ or 2.3×10^{-24} g. In a similar fashion the weight of an atom of any element can be calculated if we know its atomic weight. The dimensions of the atom can be approximated using a similar calculation.

The volume of 12 g of $^{12}C_6$ is equal to its mass divided by 2.3, or approximately 5.3 ml. The average space occupied by a single atom is therefore $5.3 \div 6.02 \times 10^{23}$, or approximately 9×10^{-24} ml. This number, of course, is only an approximation since it does not take into account the actual physical shape of the atom or the space between one atom and the next. In fact, more accurate methods reveal that the effective radius of carbon as graphite is $0.77°A$, which leads to a volume of approximately 2.5×10^{-25} $(0.77 \times 10^{-8})^3$ cm. But the approximation shows agreement within an order of magnitude with the observed value.

It is interesting to reflect how small this is. It helps us to realize how rapidly a particle this small can diffuse away from a particular space of interest such as a light path or how easily it can diffuse through an apparently solid material such as graphite (as in a graphite atomizer used in atomic absorption) atomizers.

c. Atomic Structure

The first significant attempt to explain the structure of atoms in terms of electrons rotating around a nucleus was made by Niels Bohr. His theory was based on Max Planck's quantum theory and the concept of Rutherford that the atom contained a nucleus. Bohr proposed that the negatively charged electrons moved around a positively charged nucleus and were attracted by an electrostatic force which was balanced by the centrifugal force of their circular movement. He assumed that for the atom to be stable, the electrons could only travel in certain restricted orbits, which he called "*stationary states.*"

Later his theory was modified. It was proposed that groups of electrons made up whole shells with the nucleus at the center. Each shell was assigned a principal quantum number, n, so that the first shell $n = 1$, second shell $n = 2$, and so on. It was then further proposed that the orbits of the electrons were not circular but elliptical. An ellipse is the path of a point such that the sum of its distances from two fixed points is constant. Two points are therefore needed to define the ellipse. The first point is the nucleus. The second point is called the azimuthal quantum number and is designated by the letter l. Experimental evidence showed that the orbits were split into sublevels by a magnetic field, and this number is defined as the magnetic quantum number m. Finally, the direction of spinning of the electron around its own axis could be clockwise or counterclockwise providing a spin quantum number s. The orbit of the electron was therefore defined by the four quantum numbers n, l, m, s, which were the principle quantum number n, the azimuthal number l, magnetic number m, and spin quantum number s. These numbers are related to each other as follows. n is the principal quantum number, and the other numbers define the sublevels of that shell. If n is the principal quantum number, the azimuthal quantum number is any number up to $n - 1$. m, the magnetic quantum number, is any number up to $\pm l$, and s is one of the two numbers $+\frac{1}{2}$ or $-\frac{1}{2}$. So if $n = 3$, the l may be 0, 1, or 2, m may be -2, -1, 0, $+1$, or $+2$, and $s = +\frac{1}{2}$ or $-\frac{1}{2}$.

Later Louis deBroglie proposed that electrons behaved as waves and that the lengths of the electron's path around a nucleus must be a whole number of waves for the atom to be stable. Erwin Schrödinger then developed the Schrödinger wave equation, which is the basis for modern wave mechanics. Solutions of the Schrödinger wave equation define the energy of the electrons. Four numbers must be defined in the solution of the wave equation. These four numbers are identical to the four quantum numbers n, l, m, s of the classical theory.

A further refinement was developed by Wolfgang Pauli, who stated that no two electrons in a given atom can have the same set of quantum numbers.

In 1913 Bohr postulated that in an atom electrons revolved around the nucleus and must move in one of a number of orbitals specified by quantum theory. These were called stationary states. Moving in any one of these stationary states the electrons do not radiate energy as classical electrodynamics would have us expect. Radiation is only emitted when the electron jumps from one stationary state to another. Bohr's second postulate was that the energy of the radiation and therefore the frequency of the radiation depends on the energy difference between the two stationary states.

Modern spectroscopy has been built on these two postulates. Knowing the frequency of the many lines emitted by an excited atom, physicists were able over many years to deduce the energy levels of the stationary states. This enabled them to account for the many emission lines observed from excited atoms and in some cases to predict where lines should appear.

This vector model of an atom provides a model which can be visualized, but it is not as comprehensive as the wave mechanical model.

Unfortunately, the wave mechanical model is itself difficult to visualize, and in practice the equations have only been applied to understanding some of the simpler problems since they are so difficult to solve.

d. Electron Configuration

It is common to designate all electrons with principal quantum number $n = 1$ as being in the K shell. Similarly, all electrons with principal quantum number $n = 2$ are defined as being in the L shell, $n = 3$, the M shell, $n = 4$, the N shell, $n = 5$, the O shell, etc. throughout the Periodic Table. Further, it has been found that there is a relationship between the principal quantum number and the number of electrons in that shell, as shown in Table 1.1.

The azimuthal quantum numbers are designated by the letters s p d f. These letters are remnants of the older theory where emission lines were characterized as sharp (s), principle (p), diffuse (d), and fine spectral (s) lines. s has an azimuthal quantum number equal to 0, $p = 1$, $d = 2$, $f = 3$, etc. The term $2p^5$ designates 5 electrons with a principle quantum number 2 and azimuthal quantum number 1.

A list of the designated configurations of the elements of the Periodic Table is shown in Table 1.2. As we go through the Periodic Table from one element to the next, we systematically increase the charge on the nucleus (number of protons, i.e., atomic number) and the number of electrons. The electrons first fill the orbitals with the *lowest energy* leaving the upper orbitals empty. For example, the order of filling for element atomic number 49 (indium) would be $1s^2$, $2s^2$, $2p^6$, $3s^2$, $3p^6$, $4s^2$, $3d^{10}$, $4p^6$, $5s^1$, $4d^{10}$, and $5p^1$. The upper orbitals such as $5p^2$, $5d^1$, etc. would remain empty. Note that $5s^2$ has lower energy than $4d^{10}$, although the principle quantum number is higher. The overlap of shells is not uncommon. The relationship between quantum numbers and orbitals is shown in Table 1.3.

In order to designate the electronic structure of a particular element, it is too repetitive to quote all the filled energy levels since the inner shells will be the same each time. It is therefore common only to mention the last pertinent electrons, such as the $5p^1$ electron in the case of indium, i.e., an electron with principal quantum number 5 and azimuthal quantum number p (i.e., 1). Element 50 would be $5p^2$, i.e., with a second electron with principal quantum number 5 and azimuthal quantum number p (i.e., 1).

When these electrons are arranged in order, it is easier to understand the Periodic Table and the relationship between the chemical properties of the elements.

Table 1.1 Number of Electrons in Different Shells

n	1	2	3	4
Shell	K	L	M	N
Number of electrons	2	8	18	18

Table 1.2 Electron Configuration of the Elements

Shells 72–92 (Hf–U):

n = 1		K	L	M	N	O (5)			P (6)			Q (7)
		1	2	3	4	s	p	d	s	p	d	s
72	Hf	2	8	18	32	2	6	2	2			
73	Ta	2	8	18	32	2	6	3	2			
74	W	2	8	18	32	2	6	4	2			
75	Re	2	8	18	32	2	6	5	2			
76	Os	2	8	18	32	2	6	6	2			
77	Ir	2	8	18	32	2	6	7	2			
78	Pt	2	8	18	32	2	6	9	1			
79	Au	2	8	18	32	2	6	10	1			
80	Hg	2	8	18	32	2	6	10	2			
81	Tl	2	8	18	32	2	6	10	2	1		
82	Pb	2	8	18	32	2	6	10	2	2		
83	Bi	2	8	18	32	2	6	10	2	3		
84	Po	2	8	18	32	2	6	10	2	4		
85	—	2	8	18	32	2	6	10	2	5		
86	Rn	2	8	18	32	2	6	10	2	6		
87	—	2	8	18	32	2	6	10	2	6		1
88	Ra	2	8	18	32	2	6	10	2	6		2
89	Ac	2	8	18	32	2	6	10	2	6	1	2
90	Th	2	8	18	32	2	6	10	2	6	2	2
91	Pa	2	8	18	32	2	6	10	2	6	3	2
92	U	2	8	18	32	2	6	10	2	6	4	2

Shells 37–71 (Rb–Lu):

n = 1		K	L		M			N (4)				O (5)				P (6)
		1	2 s	2 p	3 s	3 p	3 d	4 s	4 p	4 d	4 f	5 s	5 p	5 d	5 f	6 s
37	Rb	2	2	6	2	6	10	2	6			1				
38	Sr	2	2	6	2	6	10	2	6			2				
39	Y	2	2	6	2	6	10	2	6	1		2				
40	Zr	2	2	6	2	6	10	2	6	2		2				
41	Nb	2	2	6	2	6	10	2	6	4		1				
42	Mo	2	2	6	2	6	10	2	6	5		1				
43	Ma	2	2	6	2	6	10	2	6	6		1				
44	Ru	2	2	6	2	6	10	2	6	7		1				
45	Rh	2	2	6	2	6	10	2	6	8		1				
46	Pd	2	2	6	2	6	10	2	6	10						
47	Ag	2	2	6	2	6	10	2	6	10		1				
48	Cd	2	2	6	2	6	10	2	6	10		2				
49	In	2	2	6	2	6	10	2	6	10		2	1			
50	Sn	2	2	6	2	6	10	2	6	10		2	2			
51	Sb	2	2	6	2	6	10	2	6	10		2	3			
52	Te	2	2	6	2	6	10	2	6	10		2	4			
53	I	2	2	6	2	6	10	2	6	10		2	5			
54	Xe	2	2	6	2	6	10	2	6	10		2	6			
55	Cs	2	2	6	2	6	10	2	6	10		2	6			1
56	Ba	2	2	6	2	6	10	2	6	10		2	6			2
57	La	2	2	6	2	6	10	2	6	10		2	6	1		2
58	Ce	2	2	6	2	6	10	2	6	10	1	2	6	1		2
59	Pr	2	2	6	2	6	10	2	6	10	2	2	6	1		2
60	Nd	2	2	6	2	6	10	2	6	10	3	2	6	1		2
61	Il	2	2	6	2	6	10	2	6	10	4	2	6	1		2
62	Sm	2	2	6	2	6	10	2	6	10	5	2	6	1		2
63	Eu	2	2	6	2	6	10	2	6	10	6	2	6	1		2
64	Gd	2	2	6	2	6	10	2	6	10	7	2	6	1		2
65	Tb	2	2	6	2	6	10	2	6	10	8	2	6	1		2
66	Dy	2	2	6	2	6	10	2	6	10	9	2	6	1		2
67	Ho	2	2	6	2	6	10	2	6	10	10	2	6	1		2
68	Er	2	2	6	2	6	10	2	6	10	11	2	6	1		2
69	Tm	2	2	6	2	6	10	2	6	10	12	2	6	1		2
70	Yb	2	2	6	2	6	10	2	6	10	13	2	6	1		2
71	Lu	2	2	6	2	6	10	2	6	10	14	2	6	1		2

Shells 1–36 (H–Kr):

n = 1		K	L		M			N (4)	
		1 s	2 s	2 p	3 s	3 p	3 d	4 s	4 p
1	H	1							
2	He	2							
3	Li	2	1						
4	Be	2	2						
5	B	2	2	1					
6	C	2	2	2					
7	N	2	2	3					
8	O	2	2	4					
9	F	2	2	5					
10	Ne	2	2	6					
11	Na	2	2	6	1				
12	Mg	2	2	6	2				
13	Al	2	2	6	2	1			
14	Si	2	2	6	2	2			
15	P	2	2	6	2	3			
16	S	2	2	6	2	4			
17	Cl	2	2	6	2	5			
18	A	2	2	6	2	6			
19	K	2	2	6	2	6		1	
20	Ca	2	2	6	2	6		2	
21	Sc	2	2	6	2	6	1	2	
22	Ti	2	2	6	2	6	2	2	
23	V	2	2	6	2	6	3	2	
24	Cr	2	2	6	2	6	5	1	
25	Mn	2	2	6	2	6	5	2	
26	Fe	2	2	6	2	6	6	2	
27	Co	2	2	6	2	6	7	2	
28	Ni	2	2	6	2	6	8	2	
29	Cu	2	2	6	2	6	10	1	
30	Zn	2	2	6	2	6	10	2	
31	Ga	2	2	6	2	6	10	2	1
32	Ge	2	2	6	2	6	10	2	2
33	As	2	2	6	2	6	10	2	3
34	Se	2	2	6	2	6	10	2	4
35	Br	2	2	6	2	6	10	2	5
36	Kr	2	2	6	2	6	10	2	6

Table 1.3 Electronic Orbital Assignments

Shell	Quantum Number				
	n	l	m_1	S	
K	1	0	0	$\frac{1}{2}$	$1s^2$
	1	0	0	$-\frac{1}{2}$	$1s^2$
L	2	0	0	$\frac{1}{2}$	
	2	0	0	$-\frac{1}{2}$	$2s^2$
L	2	1	−1	$\frac{1}{2}$	
	2	1	−1	$-\frac{1}{2}$	
	2	1	0	$\frac{1}{2}$	
	2	1	0	$-\frac{1}{2}$	$2p^6$
	2	1	1	$\frac{1}{2}$	
	2	1	1	$-\frac{1}{2}$	
M	3	0	0	$\frac{1}{2}$	
	3	0	0	$-\frac{1}{2}$	$3s^2$
	3	1	−1	$\frac{1}{2}$	
	3	1	−1	$-\frac{1}{2}$	
M	3	1	0	$\frac{1}{2}$	
	3	1	0	$-\frac{1}{2}$	$3p^6$
	3	1	1	$\frac{1}{2}$	
	3	1	1	$-\frac{1}{2}$	
	3	2	−2	$\frac{1}{2}$	
	3	2	−2	$-\frac{1}{2}$	
	3	2	−1	$\frac{1}{2}$	
	3	2	−1	$-\frac{1}{2}$	
M	3	2	0	$\frac{1}{2}$	
	3	2	0	$-\frac{1}{2}$	$3d^{10}$
	3	2	1	$\frac{1}{2}$	
	3	2	1	$-\frac{1}{2}$	
	3	2	2	$\frac{1}{2}$	
	3	2	2	$-\frac{1}{2}$	

But our concern is mostly with spectroscopy. It is concerned mainly with the transition of electrons between these orbitals.

e. Transition of Electrons

As we have mentioned, the electrons fill the orbitals with the lowest energy in sequence. An unexcited atom will have all its electrons in the lowest possible energy states.

However, if energy is supplied to the atom, it is possible to move an electron from its lowest permitted energy state to a higher empty orbital. In the case of element 49, we could move the $5p^1$ electron to any empty $5d$ or $5f$ orbital or any other higher empty orbital. In the process the atom has become *excited*.

The fields of atomic absorption and atomic emission are concerned intimately with the transitions of electrons from one orbital to another. When the electron is in the lowest energy level, the atom is in the *ground state*. When it is in a higher orbital, the atom is in an *excited state*. The transition from the ground state to the lowest excited state is the easiest to effect. The radiation absorbed is the *resonance line*. In the case of atomic absorption the electron moves from the lowest energy level, i.e., $5p^1$ for element 49 to a higher orbital, e.g., $5d^1$.

In the case of atomic emission there is a transition of an electron from an upper electronic excited state to any lower excited state including the ground state itself. The processes that control these electronic transitions control the effectiveness of atomic absorption and atomic emission.

f. Boltzmann Distribution

The electron involved in the transitions in excitation of the atom is always the electron with the least energy, i.e., the last one in the designated list of orbitals as shown in Table 1.2. Chemically this is the valence electron. An electron can be promoted from the ground state to an upper excited state of an atom. These upper states can be populated by thermally heating a system of atoms. If a whole population of atoms is heated, the actual number of electrons in any particular orbital is given by the Maxwell-Boltzmann equation. This states that if there are N_1 atoms in an excited state and N_0 atoms in the ground state, then

$$\frac{N_1}{N_0} = \frac{g_0}{g_1} e^{-E/kT} \qquad (1.1)$$

where

E = energy difference between the ground state and the excited state involved
T = the temperature of the atom population
k = Boltzmann constant
$\dfrac{g_0}{g_1}$ = statistical weights in the energy states 0 and 1

There are several considerations in this law. First and most important, it assumes that the population of atoms is in thermal equilibrium and that they have attained a constant uniform temperature. Second, the law is universal and does not only apply to the excitation of electrons as in the case considered; it also applies to vibrational energy, rotational energy, etc.

This law is extremely important because it allows us to calculate how many atoms in a system will be excited by placing it in a flame, a thermally heated furnace, an electrical discharge, or a plasma.

The Boltzmann distribution defines the *temperature* as the *thermal temperature* of the system, which of course is related to the translational energy of the atoms or molecules involved. However it is not uncommon for the electronic emission intensity to reflect populations of excited atoms considerably higher than would be anticipated from the measured thermal temperature of the system. If excitation of electrons is involved, we can talk of the *electronic temperature* of a system. That is, we can say that the population of excited electrons is acting *as though* its thermal temperature is the electronic temperature and is such as to provide the population of excited atoms observed. We may also have a "vibrational temperature" or "rotational temperature." These temperatures satisfy the Boltzmann distribution but may not be equal to each other and may not be thermal temperatures at all.

The Boltzmann distribution also assumes that the energy levels are nondegenerate, i.e., they are single energy levels. However, often when the atoms are put in a strong magnetic field, these energy levels may split into a number of sublevels $(2Ji+1)$ Nl, where Nl is the population of each of the nondegenerate sublevels.

2. Radiation

We are all familiar with radiation as it is emitted from the sun, electric lights, fires, etc., but the actual nature of radiation is still somewhat of a mystery. A number of classical experiments have been performed which give contradictory results. Some experiments suggest that light is a wave form of energy. These experiments include the observation of interference fringes or the diffraction of radiation by crystals or gratings. Equally conclusive experiments indicate that light is a particle. These experiments include Compton scattering, Einstein's photoelectric effect, and Raleigh's detection of weak gamma emission at long distances. These contradictory results indicate that light sometimes behaves as a wave and sometimes as a particle. A mathematical treatment of the properties of light can ignore this dilemma and allows us to "understand" the interaction of radiation and matter based on the mathematical model as opposed to the physical model.

Radiation can be characterized by three factors: the velocity, the wavelength and the frequency. These are related by the equation

$$\nu = \frac{c}{\lambda} \tag{1.2}$$

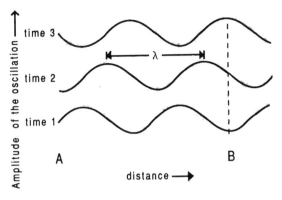

Figure 1.1 Radiation waves starting at times 1, 2, 3. Phase of light reaching point B depends on the initial phase, the wavelength, and the distance between A and B if c is constant.

where

ν = frequency in cycles per second
λ = wavelength in centimeters
c = velocity of light in centimeters per second

But c is constant, therefore when λ increases, ν decreases and vice versa. The equation can be understood by looking at Fig. 1.1. Radiation emitting from a point A will pass point B. The velocity at which it passes between these two points is c, the speed of light. The number of waves that pass point B per unit of time is the frequency, and the distance between the crests of the waves is equal to λ, the wavelength. The velocity of radiation in a vacuum is equal to 2.997925×10^{10} cm per second, and the Special Case of Relativity states that this velocity is independent of movement of the radiation source and/or the detector.

The wavelength of light determines what kind of light we consider it to be. For example, the wavelength of visible light is between 450 and 750 nm.

If light is considered a particle, each particle is called a photon. If it is considered a wave, there is no distinct photon involved. However Max Planck provided a model which insists on the existence of both of these two pictures of radiation. Prior to Planck it was believed that oscillating particles would generate energy on a continuous wavelength basis. This did not fit the experimental work of Planck, who observed the energy distribution from the hot body as a function of wavelength of radiation. He proposed therefore that radiation could only be emitted or absorbed in bundles containing discrete amounts of energy. This energy is defined by the equation

$$E = h\nu \tag{1.3}$$

where

E = the energy of the bundle (photon)
h = a universal constant (Planck's constant) and is equal to 6.24×10^{-27} erg per second
ν = the frequency of radiation

These energy bundles are called quanta, and Eq. (1.1) provided the basis for the quantum theory. It states that each pulse of energy is a photon or quantum and has a precise amount of energy, given by Eq. (1.3).

This equation can be combined with Einstein's relationship equating energy E to mass, and the two equations $E = h\nu$ and $E = mc^2$ provide the relationship:

$$h\nu = mc^2 \quad \text{or} \quad m = \frac{h\nu}{c^2}$$

but

$$\nu = \frac{c}{\lambda} \quad \text{hence} \quad m = \frac{h}{c\lambda}$$

The implication of this equation is that the photon has mass and that the mass is a function of its wavelength. A physical model of a particle photon with an associated wavelength has been proposed but not confirmed (3).

We have artificially defined different ranges of wavelengths according to experimental work and observations which have been carried out with these wavelengths. This is in essence the electromagnetic spectrum and is shown in Table 1.4.

It should also be noted that Eq. (1.1) can be rewritten, since

$$E = h\nu$$
$$= \frac{c}{\nu}$$

but

$$E = \frac{hc}{\lambda} \tag{1.4}$$

Further, as frequency ν increases, the energy of the photon increases and, conversely, as the wavelength increases, the energy of the photon decreases. Some physical *properties* of radiation are shown in Table 1.4.

The relationship between the type of radiation and the associated analytical field is shown in Table 1.5.

Table 1.4 The Electromagnetic Spectrum

Radiation	Approximate wavelength cm	μm	Commonly used dimension	Frequency (Hz)	Energy per photon (eV)
Radio	1×10^6	10^{10}	10KM	3×10^4	3×10^{-10}
	100	10^6	1M	3×10^8	3×10^{-8}
Radar UHF	10^2	10^6	M	3×10^8	2×10^{-6}
	1	10^4	cm	3×10^{10}	2×10^{-4}
Microwave	10^2	10^6	100 cm	3×10^8	2×10^{-6}
	10^{-2}	10^2	1mm	3×10^{12}	2×10^{-2}
IR	10^{-2}	100	100μ	7.5×10^{12}	2×10^{-2}
	7×10^{-5}	0.7	.7 μm or 700 nm	4.3×10^{14}	1.2
Visible	7×10^{-5}		700 nm	4.3×10^{14}	1.2
	4.5×10^{-5}		450 nm	6.6×10^{14}	2.7
UV	4.5×10^{-5}		450 nm	6.6×10^{14}	2.7
	2×10^{-5}		200 nm	1.5×10^{15}	6.2
Vac. UV	2×10^{-5}		200 nm	1.5×10^{15}	6.2
	1×10^{-6}		10 nm, 100 A	3×10^{16}	1.2×10^3
X-ray	10^{-6}		100 A	3×10^{16}	1.2×10^3
	10^{-9}		0.1 A	3×20^{19}	1.2×10^5
Cosmic	10^{-9}		0.1 A	3×10^{19}	1.2×10^5
	10^{-13}		0.1mA	3×10^{23}	1.2×10^9

a. Atomic Spectra

The interaction of radiation and matter is the basis of spectroscopy. In essence two things can happen. Radiation can be either *absorbed* by matter or *emitted* by matter. The measurement of these phenomenon provides the basis of both absorption and emission spectroscopy.

The absorption spectra in atomic spectroscopy are relatively simple compared to those in molecular spectroscopy. The energy levels permitted in atoms are the energies of the electronic orbitals described earlier. The energy differences are the difference between these energies. If an electron goes from a ground state E_0 to an excited state E_1, there is a difference of energy $E_0 - E_1$. In order to undergo transition it must absorb a photon with the same energy, i.e., $E_0, - E_1 = E = h\nu$, and $h\nu$ is the energy of the absorbed photon. This absorption of energy is the basis of atomic absorption spectroscopy.

Similarly an excited atom will have an electron in an upper excited orbital. The electron descends to a lower orbital of lower energy. The change in energy is equal to $E_4 - E_3$, where E_4 and E_3 are the energies of the upper orbitals concerned. In the process a photon of energy is emitted. The energy of this photon must equal E',

Table 1.5 Relationship Between Radiation and the Associated Analytical Field

Radiation	Physical effect	Analytical field
Radio	Nuclear spin change	NMR
Microwave	Molecule rotates	Microwave spectroscopy
IR	Molecule vibrated	IR
Visible	Electronic excitation	
UV	electron is excited	Atomic absorption
		Colorimetry
	excited electron	UV absorption
	relaxes	Emission spectrography
		Plasma emission
		Flame photometry
		UV fluorescence
		UV phosphorescence
Vac UV	Outer shell	Vac UV
	nonvalence electron excited	absorption
X-ray	Inner shell electron	X-Ray absorption
	displaced	X-Ray diffraction
		X-Ray fluorescence
		ESCA
		Auger
Gamma rays	Nucleus changes	Nuclear
		science
Cosmic rays		Radiation from space

where E' is the difference in energy between E_3 and E_4, i.e., $E' = h\nu = E_4 - E_3$ and it is specific of the transition E_4 to E_3. This is the basis for emission spectroscopy, plasma emission, and flame photometry.

The emission spectra of atoms have been observed and studied for many years. By a slow methodical mathematical process of trial and error, the emission lines have been assigned to transitions between particular energy levels and the energy of those levels has been calculated. Diagrams have been developed for each element assigning the energies of each orbital and the wavelength of the transition between those orbitals. These are called Grotrian diagrams. They are used extensively in atomic absorption and atomic emission. They are illustrated below (4).

As we shall see later, in atomic absorption only transitions involving the ground state are involved. In atomic emission, theoretically, any transition can be used, but experience has enabled us to make judicious selections.

3. Formation of Free Atoms

It is of practical importance to be aware of the energies required to break chemical bonds because this represents the energy which must be supplied by the atomizer to liberate atoms bound in a molecule to generate free atoms.

Atoms group together and form molecules. The atoms are held to each other by chemical bonds. There are several types of chemical bonds, the most important being ionic, coordinate, and covalent. In ionic bonds the elements concerned are ionized positively if they are metals and negatively if they are nonmetals. To the first approximation they are held together by the attraction of their opposite charges, but in practice it is found that the bond often has covalent character as well as ionic character. Table 1.6 indicates some typical distributions of bond character of the acids of the halides and the bond energies.

In covalent and coordinate bonds there is overlap of an s orbital of one atom with an s orbital of another atom on a p orbital of one atom and p orbital of the second atom. These form sigma (σ) or pi (π) bonds, the sigma indicating that it is a single bond formed between the two elements concerned. If a double bond is formed in which the atoms share more than one pair of electrons, this is signified by a π bond, in which case the double bond would include a single bond and a π bond. The energy required to break some of these bonds is shown in Table 1.7.

An atomizer such as a flame, carbon atomizer, or plasma must break these bonds to free the atoms. The energy required greatly controls the efficiency at which free atoms are produced from a sample. Consequently the efficiency of a flame atomizer to generate free atoms of a given element such as iron may be different if the iron is ionized (such as a solution of iron chloride) or complexed (such as a solution of iron EDTA). Such samples, even with the same iron concentration, would give different analytical signals. It provides the basis for understanding chemical interference (see page 103), which is a major problem in atomic absorption and atomic emission spectroscopy. The phenomenon explains why cool atomizers are not as efficient as hot atomizers for many elements.

Table 1.6   Bond Character and Strength of Halo Acids

Bond	Electronegativity difference	Percent ionic character	Percent covalent character	Bond strength (kcal/mole)
H-F	1.8	55	45	135
H-Cl	0.9	19	81	103
H-Br	0.7	12	88	88
H-I	0.4	4	96	71

Table 1.7 Bond Energies (kcal/mole at 25°C)

Diatomic Molecules:

H–H	104	F–F	37	H–F	135
O=O	119	Cl–Cl	58	H–Cl	103
N N	226	Br–Br	46	H–Br	88
C=O	256	I–I	36	H–I	71

Polyatomic Molecules:

C–H	99	C–C	83	C–F	116
N–H	93	C=C	146	C–Cl	81
O–H	110	C C	200	C–Br	68
S–H	83	C–N	73	C–I	51
P–H	76	C=N	147	C–S	65
N–N	39	C N	213	C=S	128
N=N	100	C–O	86	N–F	65
O–O	35	C=P(CO$_2$)	192	N–Cl	46
S–S	54	C=O(HCHO)	166	O–F	45
N–O	53	C=O(RCHO)	176	O–Cl	52
N=O	145	C=O(RCOR′)	179	O–Br	48

C. COMPARISON OF ABSORPTION AND EMISSION

The absorption of energy by atoms follows well-known physical laws and appears to be predictable, thus providing us with a basis for quantitative analytical chemistry. The radiant energy absorbed by the atoms is generally in the form of very narrow absorption lines with wavelengths in the visible or ultraviolet region of the radiant energy spectrum. During the absorption process the outer valence electrons of the atoms jump to a higher orbital and the atom is said to become excited (electronic excitation). To a first approximation there is a very simple relationship between the populations of excited and unexcited atoms, and therefore between atomic absorption and atomic emission spectroscopy. This simple relationship is shown in Fig. 1.2, which simplistically illustrates the equilibrium between excited atoms and unexcited atoms plus a photon. The unexcited atom is said to be in the *ground state.*

The generation of a lower energy state plus a photon from a higher energy state (excited) atom is the basis of emission spectroscopy. In this technique we measure the wavelength and number of photons generated in the emission process. The reverse process of absorbing a photon by a ground-state atom to become an excited atom is the basis of atomic absorption spectroscopy. Except under special circumstances, we do not measure the "number of photons emitted or absorbed";

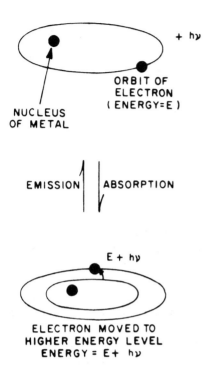

Figure 1.2   The relationship between atomic absorption and atomic emission spectroscopy.

rather we measure the "intensity of light" emitted in the case of emission or the percentage of light absorbed in the case of absorption and relate this through a calibration curve to the metal concentration in the original sample.

In any atom there are numerous permitted energy levels and numerous transitions permitted between different excited energy levels and between excited energy levels and the ground state (see page 8). The existence of numerous upper excited states points out that the equilibrium between ground-state atoms and excited atoms is not as quite as simple as in Fig. 1.2. However, for the sake of clarity and for the purpose of illustrating the principles of emission and absorption, the upper energy states will be disregarded at this point, and we shall limit our discussion to the ground state and the first excited state of the atom.

Figure 1.2 helps to reveal the gross effects of variable conditions of a system of excited and unexcited atoms. For example, any change that increases the total population of atoms generated from a given sample will increase both the number of excited atoms and the number of unexcited atoms. This in turn leads to an increase in both the emission signal and the absorption signal. The use of highly

Table 1.8 The Relationship Between Temperature, Excitation Wavelength, and the Number of Atoms in a Unit Atom Population

Excitation wavelength (nm)	Number of excited atoms per unit population at		Enhancement caused by 500 K temperature increase
	3000 K	3500 K	
200	10^{-10}	$10^{-8.6}$	30
300	$10^{-6.7}$	$10^{-5.7}$	10
600	$10^{-3.3}$	$10^{-3.0}$	2.3

combustible and volatile solvents, high atomizer temperatures, and organometallic compounds all tend to increase the total number of free atoms formed in a flame atomizer from a given sample concentration. Because there are numerous excited states available to the atom, however, the increase in emission intensity from any particular level does not bear a simple relationship to the increase in atomization efficiency. On the other hand, with increased atomization efficiency the degree of absorption increases almost linearly.

In contrast to these circumstances, if a particular condition favors the formation of either ground-state or excited atoms (but not both), *for a constant atom population* we shall get an increase of either emission and a drop in absorption or vice versa. An example of this is an increase in flame temperature. For a *fixed population* of free atoms, if the atomizer temperature is increased, there will be an increase in the number of excited atoms and a corresponding decrease in the number of unexcited atoms (Table 1.8). Thus, we can see that increasing atomizer temperature will increase both the efficiency of atomization and therefore emission and absorption; but it will also change the balance of the number of atoms and produce more excited atoms. It can be seen, therefore, that the effect of temperature is not the same for emission spectroscopy as for atomic absorption spectroscopy. The same can be said for other variables, and we shall try to understand these effects by considering each of the variables in turn and how they effect emission and absorption.

One of the most basic and most important relationships in spectroscopy was given in Eq. (1.3).

$$E = h\nu$$

where

E = energy difference between two energy levels in an atom (or molecule) between which transition occurs

h = Planck's constant
v = frequency of radiation

$$= \frac{c}{\lambda}$$

where

c = speed of light
λ = wavelength of radiation

The importance of Eq. (1.2) gives the frequency (and hence the wavelength) of radiation associated with an energy shift from one energy state to another in an atom or molecular system. We shall confine our discussions to atomic energy levels inasmuch as we are principally concerned with atomic spectroscopy.

Eq. (1.3) tells us that the energy change E in the atom must equal the energy hv of the associated photon.

The quantum theory tells us that only certain energy levels are permitted in atoms (or molecules). This means that the atom can exist for extended periods only at permitted energy levels; extended existence at all other energy levels is forbidden and can be ignored in this branch of spectroscopy.

Since only certain energy levels are permitted, the energy difference between these levels is well defined and thus only certain wavelengths of radiation can be emitted or absorbed by an atom. Consequently, the atomic emission spectrum of a particular element is characteristic of that element. Similarly only certain radiation frequencies can be *absorbed* by atoms of a particular element. These two properties provide us with the specificity necessary to make atomic absorption and emission spectroscopy useful, reproducible analytical tools. If an atom such as sodium is found to absorb a wavelength of 589 nm, then we can be assured that it will always absorb at 589 nm because the energy levels associated with the absorption of this wavelength of energy are physical properties of sodium.

If a system of atoms is held at an elevated temperature, a number of the atoms will exist in the upper excited states. The intensity of emission depends on the number of atoms in these excited states. This relationship governed by the Boltzmann distribution is:

$$\frac{N_1}{N_0} = \frac{g_1}{g_0} e^{-E/kT}$$

where

$\dfrac{g_1}{g_0}$ = a priori probabilities or statistical weights of the atoms in state 1 and state 0

N_1 = Number of atoms in the excited state

N_0 = Number of atoms in the ground state
E_2 = energy difference $E_1 - E_0$
T = temperature of the system
k = Boltzmann constant

Emission intensity S is given by the relationship

$$S = \frac{N_1 E}{\tau} \qquad (1.5)$$

where

τ = lifetime in the excited state

From Eq. (1.1), therefore,

$$S = \frac{N_0 E}{\tau} \frac{g_1}{g_0} e^{-E/kT} \qquad (1.6)$$

Some specific examples are given in Table 1.9.

Further, the intensity of the emitted radiation is proportional to the number of excited atoms. For a given total number of atoms, the number of excited atoms is a function of the temperature and the excitation energy E. As illustrated by the Boltzmann distribution, the energy of excitation can be found by measuring the frequency of the radiation emitted as indicated by Eq. (1.3).

The Boltzmann distribution can be used to advantage if we are careful to remember that there are differences between translational temperatures and other temperatures such as "electronic" temperature, "vibrational" temperatures, "rotational" temperatures, etc. It is not uncommon for the system of atoms or molecules

Table 1.9 Effect of T and λ on Excited Atom Population

	Resonance Line	gm/gn	ΔE (in eV)	No. excited atoms, N_1/No. unexcited atoms, N_0	
				2000 K	3000 K
Cs	851.1 nm	2	1.45	4.4×10^{-4}	7.2×10^{-3}
Na	589.0 nm	2	2.10	9.9×10^{-6}	5.9×10^{-4}
Ca	422.7 nm	3	2.93	1.2×10^{-7}	3.7×10^{-5}
Fe	372.0 nm		3.33	2.3×10^{-9}	1.3×10^{-6}
Cu	324.8 nm	2	3.82	4.8×10^{-10}	3.7×10^{-7}
Mg	285.2 nm	3	4.35	3.4×10^{-11}	1.5×10^{-7}
Zn	213.9 nm	3	5.80	7.5×10^{-15}	5.5×10^{-10}

to exist in several different temperature states at the same time. For example, the electronic temperature and the vibrational temperature may be quite different. This merely means that the effective Boltzmann temperature is that which is calculated from the emission intensity and may be different based on radiation emitted as a result of vibrational and electronic transitions.

In contrast, the degree of absorption in atomic absorption is given by

$$\int_0^\infty k\nu \, d\nu = \frac{\pi e^2}{mc} N_0 f \tag{1.7}$$

where

k = absorption coefficient at frequency ν
e = charge of an electron
m = mass of an electron
N_0 = number of absorbing atoms at energy level 1 (ground state)
c = speed of light
f = oscillator strength of the absorption line

The *oscillator strength* of a line is measured as the probability of a transition between the ground state and the excited state involved. In emission spectrography the oscillator strength is related directly to the intensity of emission. If one emission line is twice as strong as another emission line from the same spectrum, then the oscillator strengths of these two lines are in the ratio 2:1. The greater the oscillator strength, the more likely a transition will take place. The oscillator strength can be defined mathematically as

$$f = \frac{mc}{8\pi^2 e^2} \frac{g_1}{g_2} \lambda_2^1 A_1^2 \tag{1.8}$$

where

f = oscillator strength
λ = wavelength of the resonance line between levels 1 and 2
A_1^2 = Einstein's coefficient of spontaneous emission
$\dfrac{g_1}{g_2}$ = statistical weights of atoms in states g_1 and g_2

It can be readily seen that f, the oscillator strength, is a physical property of a particular element and is therefore not a variable under normal conditions.

If we look back at Eq. (1.7), we see that the degree of absorption is equal to the constants e, m, c, and f multiplied by the total number of atoms in the light path. This relationship demonstrates one of the fundamental differences between atomic emission and atomic absorption. The temperature, T, and the energy of transition, E, are not part of the mathematical relationship governing the degree of absorption. Hence, to a first approximation the total degree of absorption is equal to a constant times the number of free atoms in the light path and is independent of temperature and the energy of transition. This can be represented as:

$$\text{Total absorption} = \text{constant} \times Nf \tag{1.9}$$

D. THE BEER-LAMBERT LAW

The absorption by atomic systems follows Beer's law, which is

$$I = I_0 e^{-a,b,c} \tag{1.10}$$

or

$$\ln \frac{I}{I_0} = A = abc \tag{1.11}$$

where

$$T = \frac{I_1}{I_0}$$

A = absorbance = $-\ln T$
I_1 = amount of light emerging from a solution after absorption
I_0 = intensity of light falling on a solution before absorption
a = absorption coefficient
b = path length of the light through the sample
c = concentration of the solution $\leq$ 0.1 M

The Beer-Lambert law is one of the most fundamental laws relating the degree of absorption by a solution to the concentration of its components.

Although the basic principles of the Beer-Lambert law apply to atomic absorption spectroscopy, in practice it is not feasible to use this relationship in the same way it is used for the measurement of analyte concentrations in a solution. This is because solutions are invariably homogeneous, and molecular concentrations are therefore constant throughout the sample absorption light path. However, in a system of free atoms (e.g., in a flame), the concentration of free atoms is not constant throughout the absorption light path. Hence the Beer-Lambert law cannot be used directly to determine "*the concentration*" of an atom generated from sample solution. Instead it is necessary to use Eq. (1.9), relating the degree of absorption with the total number of atoms in the light path.

As will be discussed later, the free atoms in the light path at any particular moment are in a state of dynamic equilibrium with the sample solution and with the products of combustion. The number of free atoms is in turn proportional to the concentration of the metal being determined in the sample. We can therefore construct calibration curves relating the degree of absorption and the concentration of a solution. It can be seen from Table 1.8 that the number of excited atoms varies considerably with a change in temperature of the system. It can also be seen that the number of excited atoms represents a very small proportion of the total number of free atoms in that population. It can be concluded from this that the great bulk of free atoms exist in the ground state (or the unexcited state) at temperatures normally encountered in flame atomizers. These unexcited atoms do not contribute to the emission signal, but they can contribute to the degree of absorption because they are all capable of absorbing energy. Table 1.8 also reveals that even if the number of excited atoms increases greatly, the total number of unexcited atoms varies almost imperceptibly. For a given population of atoms, the ground state is the most highly populated and within experimental error is essentially independent of temperature. Since the degree of absorption is a function of the number of free atoms that can absorb energy, we can conclude that the degree of absorption is virtually independent of temperature for a given atom population. It can also be seen from Eq. (1.7) that there is no term relating the absorption wavelength to the total absorption. In contrast to emission, the total absorption is independent of the absorption wavelength.

E. SUMMARY OF THE COMPARISON OF ABSORPTION AND EMISSION TECHNIQUES

To summarize, based on the physical properties of free atoms, the following inherent relationships exist between atomic emission and atomic absorption spectroscopy.

1. The physical process of absorption by atoms is virtually independent of the temperature of the system, whereas emission intensity is very dependent on temperature.
2. Atomization efficiency, which is itself a variable, affects both emission and absorption. When flame atomizers are used, changes in flame temperature can affect the efficiency of producing atoms from a given sample and affects both absorption and emission. Therefore it directly controls N_1, the number of unexcited atoms in the light path. Such a temperature effect on atomic absorption can be readily observed experimentally.
3. Atomic absorption is independent of the wavelength being short or long. In contrast, for emission the Boltzmann distribution indicates that higher energy transitions (short wavelength) will have lower emission intensities than those produced from lower energy transitions.
4. The great majority of free atoms exist in the ground or *unexcited state* and contribute to the atomic absorption signal. In contrast, the atomic emission signal depends directly on the number of *excited atoms* in the system, which are only a small fraction of the total number of free atoms in the system.
5. It must be stated that the emission signal is measured as a small signal against

a background that approaches zero. It is possible to amplify this small signal many times and therefore compensate to some degree for the loss of absolute signal encountered in emission spectrography.

6. In atomic absorption, it is only necessary to measure the ratio of I_1 and I_0 in order to measure absorption. It is frequently easier to measure this ratio than to measure absolute quantities, particularly when the effect of interferences must be corrected for in order to obtain accurate results.

7. When carbon filament or carbon tube atomizers are used, extremely high sensitivities—on the order of 10^{-14} g—have been achieved by atomic absorption spectroscopy using relatively simple equipment. But this technique usually cannot be used for volatile elements (As, Sb, Hg, etc.) without special pretreatment.

8. In plasma emission, sensitivites rivaling those obtained with flame atomic absorption are frequently observed.

9. In plasma emission, the useful analytical range is often up to five decades, but in absorption, one or two decades is the norm.

10. Simultaneous or sequential multielement analyses are commonplace in emission spectroscopy. It is not difficult to determine 20 elements at a time within seconds.

11. Emission spectroscopy (including plasma emission) is the method of choice for the simultaneous qualitative analysis of numerous metals. It cannot be done by atomic absorption.

F. OPTICS IN SPECTROPHOTOMETRY

1. Single-Beam Optics

Perhaps the simplest spectroscopic instrument is the single-beam spectrophotometer. It consists basically of a light source, a sample container, a monochromator, and a detector/read-out system. The system is schematically identical to that shown in Fig. 1.3. For many years the most widely used instrument in routine analytical labs was the Beckman DU spectrophotometer, later modified to the DBG grating spectrophotometer. The method of operation was quite simply based on measuring the degree of absorption by the sample. The wavelength was at an absorption maximum in the spectrum of the sample to be determined. A series of standards of known concentrations was made up and the absorbance versus concentration relationship was measured and plotted, producing a curve similar to that in Fig. 2.37. This was a calibration curve. A sample was then put into the sample holder and its absorbance measured. Based on the degree of absorption measured and relative concentration shown on the calibration curve, the concentration of the sample was calculated. Sometimes it was necessary to correct for background absorption due to the solvent or other interfering bodies. This was done by running

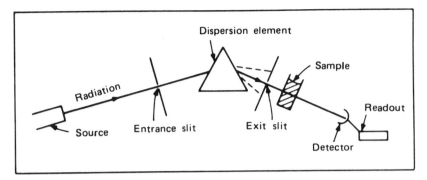

Figure 1.3 Single-beam instrument.

a blank standard in which all the ingredients of the sample except the sample element itself were added. The degree of absorption of this blank was subtracted from that of the sample, and the net absorption due to the sample was calculated and reported.

Single-beam optical systems are subject to drift, but in the hands of a trained operator, this does not generally present a problem, because he checks for drift repeatedly during operation.

Single-beam spectrophotometers have been widely used for several decades and will undoubtedly continue to be used until they are ultimately displaced by the more accurate but more expensive double-beam systems.

2. Double-Beam Optical System

The double-beam system is used extensively for spectroscopic molecular absorption studies. The individual components of the system have the same function as in the single-beam system, with one very important difference. The radiation from the radiation source is split into two beams of approximately equal intensity. One beam is termed the *reference beam*; the second, which passes through the sample, is called the *sample beam*. The two beams are then recombined and pass through the monochromator system to the detector. This is illustrated in Fig. 1.4.

As shown in Fig. 1.5a, the beam splitter may be a simple mirror plate into which a number of holes are drilled. Light is reflected by the mirror plate and passes down the sample beam path. An equal portion of light passes through the holes in the plate and forms the reference beam.

Another convenient beam splitter is a disk with opposite quadrants removed (Fig. 1.6b). The disk rotates in front of the radiation beam, and the mirrored surface

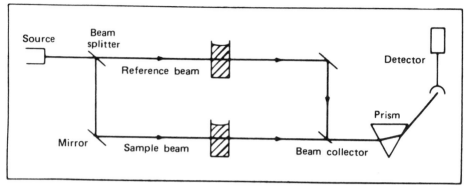

Figure 1.4   Double-beam optical system.

reflects light into the sample path. The missing quadrants permit radiation to pass down the reference beam to the detector, usually unabsorbed. When no radiation is absorbed by the sample, the two beams recombine and form a steady beam of light which falls on the detector. However, when radiation is absorbed by the sample, an alternating signal arrives at the detector (Fig. 1.6).

If there is a change in source light intensity, it affects the sample and reference beams equally. If there is no sample absorption the recombined beam will continue to give a steady signal. If there is absorption by the sample the sample beam intensity decreases, but not the reference beam. This results in an alternating

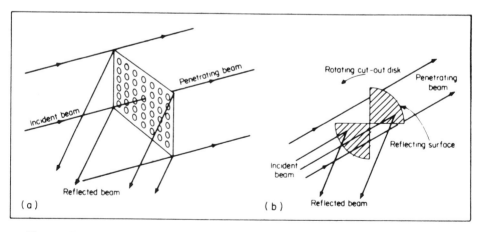

Figure 1.5   (a) Plate beam splitter and (b) disk beam splitter.

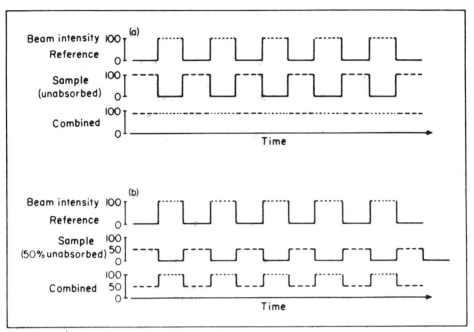

Figure 1.6 Form of the radiation intensity reaching the detector using double-beam optics: (a) no absorption and (b) the two signals reaching the detector with 50% absorption by the sample.

signal, reaching the detector, the amplitude of which is the difference between the reference beam and the sample beam. The amplitude is therefore a measure of absorption. Providing the drift is small, both beams are affected equally, and the amplitude and therefore the absorption measurement is virtually independent of source light intensity and therefore more accurate. This advantage revolutionized the studies of absorption spectroscopy.

a. Drift

Drift occurs any time there is a variation in the readout from the instrument which is caused by a failure of one or more of the components of the instrument to remain constant over a period of time. Common sources of drift are (1) variation in radiation intensity from the light source which can occur if there is a change in the line voltage to the light source, (2) variation in the readout from the detector readout system when the signal falling on that system is constant; this can also be caused by variation in line voltage, (3) any slight movement in the monochromator can change the measured radiation from the correct wavelength to a wavelength slightly off the absorption peak causing a variation in absorption. These variations can be caused by physical changes in the instrument due to expansion or cooling during daytime and nighttime or by short-term or long-term variation in the line voltage to the system.

There are two major kinds of drift: short-term drift and long-term drift. Short-term drift occurs over a very short period of time and may be the result of a rapid change in line voltage over a few seconds. This may be caused by a major user of power turning on or turning off the machine which they operate. A voltage surge can also cause drift. It is not uncommon to experience voltage surge over a short period of time as any user of a VCR will testify. Long-term drift may occur over a period of days or weeks and reflects changes in the efficiency of the light source and detector with time.

Using a double-beam system in areas of spectroscopy such as UV or infrared analysis it is quite common to put the same solvent in the reference cell as in the sample. In which case the absorption spectrum of this solvent and the reference in the sample cell cancel each other out, and the final spectrum is that of the dissolved sample and is largely not influenced by the presence of the solvent. It also corrects for absorption by such molecules as CO_2 or H_2O in the ambient atmosphere.

In atomic absorption spectroscopy the double-beam system is favored. However, the system is a *pseudo–double-beam system* because the sample cell is a flame which itself absorbs to some extent. This is particularly true at the shorter wavelengths of the ultraviolet where OH absorption bands may be of importance. However, extensive research has shown it is not feasible to use two flames in the optical system, one for the sample and one for the reference. In practice therefore only one flame is used, the reference beam remains empty, and a pseudo–double beam system is utilized.

Radiation emission systems invariably use the single-beam system. Therefore flame photometry emission spectrography and plasma emission always use a single beam system. This is because the sample atomizer is the radiation source, and splitting it into two beams achieves nothing. As a consequence there is yet to be found a convenient way to measure background using a double-beam system in an emission system. Alternate methods have been developed to correct for any emission from the source which does not emanate from the sample. This will be discussed later.

G. RELIABILITY OF RESULTS

The quantitative analysis of any particular sample should produce results that are reliable and truly representative of the sample. Unfortunately, it is virtually impossible to characterize a sample exactly, because some degree of error is always involved in the results. Under the worst conditions, this error may be gross. Not only may the results be scientifically worthless, but they may also lead to serious errors of judgment. For example, if the sample were a product from a commercial plant stream, incorrect modifications to the plant equipment might be made by the manufacturing company, and the quality of the product might be degraded as a result. The overall cost to the company might be very high. On the other hand, under the best conditions, the data would lead to correct modification and an improved product.

For analytical results to be most useful, it is important to be aware of the reliability of the results. To do this it is necessary to understand the sources of error and to be able to recognize when they can be eliminated and when they cannot. The error is the difference between the true answer and the observed answer. There are numerous sources of error, some of which are described below.

1. Types of Error

There are two principal types of errors: determinate errors and indeterminate erros. Broadly speaking, *determinate errors* are caused by faults in the analytical procedure or the equipment used in the analysis. A particular determinate error may cause the analytical results produced by the method to be always too high; another may render all results too low. Sometimes the inaccuracy produced is constant (e.g., all answers are 10% too high). This is called a *constant error*. Sometimes the inaccuracy is proportional to the true answer, giving rise to *proportional errors* such as 5% of the answer. Consequently, although the results produced by repetitive analysis of a single sample may all be too high (and inaccurate), they may nevertheless agree closely with each other, indicating high precision. It can be seen that close agreement between results (high precision) may not be an indication of the accuracy of the results.

Indeterminate errors are not constant or biased. They cause varying results, some of which may be too high and some too low. Suppose, for example, that two machines are used to measure how long it takes a bullet to reach a target. Unknown to the operator, machine A has a faulty relay switch, so that all the answers obtained with it are incorrect. Machine B, however, is in good working order. The results obtained from each machine for five shots are shown in Table 1.10. The true time it takes for the bullet to reach the target is 0.123 sec. The average of the results from machine B is 0.123 sec. Each individual reading varies slightly from this reading, but the errors average each other out. The error is indeterminate. The results from machine A vary slight from one to another (indeterminate error), but all the results are too high. This indicates a determinate error in addition to the indeterminate error.

Table 1.10 Observed Time (sec) for a Bullet to Reach a Target

Machine A (determinate and indeterminate error)	Machine B (indeterminate error)
0.130	0.120
0.133	0.122
0.132	0.123
0.136	0.124
0.134	0.120
Average 0.133	0.123

a. Determinate or Systematic Errors

Determinate errors are errors that arise because of some faulty step in the analytical process. The faulty step is repeated every time the determination is performed. If a sample is analyzed five times, the results mayall agree with each other but differ widely from the true answer. An example is given in Table 1.11. An analyst or doctor examining the analytical results might be deceived into believing that the close agreement among the answers indicates high accuracy and that the results are close to the true answer. It is the analyst's responsibility to recognize and correct for systematic errors which cause results to be consistently wrong. Two methods are commonly used to do this. One is to analyze the sample by a completely different analytical procedure that is known to involve no systematic errors. The second is to run several analyses of a standard solution of known concentration. The difference between the known (true) concentration and that

Table 1.11 Percentage of Potassium in a Patient's Blood

Observed analytical result (%)	True answer (%)
0.53	0.40
0.55	
0.54	
0.52	
0.51	
Average 0.53	0.40

Table 1.12 Percentage of Potassium in
Patients' Blood

Patient	Analytical result (%)	True answer (%)
A	0.53	0.40
B	0.75	0.62
C	0.64	0.51
D	0.43	0.30
E	0.48	0.35
Constant error of 0.13%		

obtained by analysis should reveal the error. If the results of numerous analyses are consistently high (or consistently low), then a determinate error is involved in the method. This error must be identified and a correction made before the procedure is capable of giving accurate results. Errors can arise from faulty balances, volumetric flasks, electrical dials, bent recording pens, impure chemicals, incorrect analytical procedures or techniques, operator error, low line voltages on instruments, and so on.

In the example above, the true answer was 0.40% and the observed result was 0.53%. If the doctor had repeated this procedure for numerous patients, he might have obtained the results shown in Table 1.12. It can be seen that in all cases the error was +0.13%. This indicates a constant determinate error.

On the other hand, a second doctor, working at a different hospital with different equipment and patients, might obtain the results shown in Table 1.13. Examination of these analytical results shows that they are all 25% greater than the true answer. The error is *proportional* to the answer. Such information is useful in the diagnosis of the source of the determinate error.

Table 1.13 Percentage of Potassium in
Patients' Blood

Patient	Analytical result (%)	True answer (%)
A	0.60	0.48
B	0.45	0.36
C	0.90	0.72
D	1.00	0.80
E	0.75	0.60

b. Indeterminate Errors

After all the determinate errors of an analytical procedure have been detected and eliminated, the method is still not capable of giving absolutely accurate answers. Numerous small undetectable errors may be made at each step of the procedure. Errors may be positive or negative, and the total error may be slightly too high or too low. The net error involved is an indeterminate error. All analytical procedures are subject to indeterminate error. The extent of this error can be calculated, and when known, the reliability of the method can be stated.

Indeterminate errors may arise from many sources. For example, a balance may be accurate to within 0.001 g, in which case it would discriminate between 1.015 and 1.016 but would not discriminate between two samples that weigh 1.0151 and 1.0152 g, and there would be an error in the fourth decimal place. If a balance was used accurate to within 0.000001 g, there would still be an error in the seventh decimal place. We will always have an error arising because it is outside the range of balance. Other sources of error include the limit of accuracy of volumetric equipment and the limit of readout of electrical dials or recording instruments. The significance of these effects is that a small error is always involved in each step of an analytical determination. This error may be positive or negative. If 20 measurements are made, some will be positive and some negative. If there are 10 steps, the final error is the algebraic sum of all the errors involved in each step. They may cancel each other out to some extent, but a net error may remain. If an infinite number of analyses of a single sample were carried out using this procedure, the distribution of results would be shaped like a symmetrical bell (Fig. 1.7). This shape is called Gaussian. The frequency of occurrence of any given result is represented graphically by Fig. 1.7.

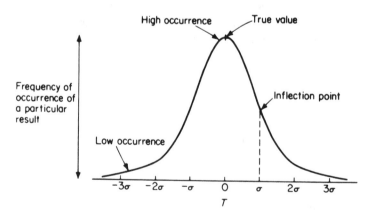

Figure 1.7 Distribution of results with indeterminate error. The range of the actual error is from low (−) to high (+). T stands for true value.

If only indeterminate errors were involved, the most frequently occurring result would be the true result (i.e., the result at the maximum of the curve) where the errors balance each other out often, however, a net positive or net negative error remains. Unfortunately, in practice it is not possible to make an infinite number of analyses of a single sample. At best, only a few analyses can be carried out, and frequently only one analysis of a particular sample is possible. We can, however, use our knowledge of statistics to determine how reliable these results should be. The basis of statistical calculations is outlined below.

2. Definitions

True value T: the true or real value.
Observed value V: the value observed by experiment.
Error E: the difference between the true value T and the observed value V (it may be positive or negative),

$$E = |V - T|$$

Mean M: the arithmetic mean of the observations, that is,

$$M = \frac{\Sigma V}{N}$$

where N is the number of observations and ΣV is the sum of all the values V. *Absolute deviation d:* the difference between the observed value V and the arithmetic mean M,

$$d = V - M$$

Deviation has no algebraic sign (+ or −); all differences are counted as positive. Other definitions include the following:

Relative deviation D: The absolute deviation d divided by the mean M,

$$D = \frac{d}{M}$$

Percentage deviation: The deviation times 100 divided by the mean,

$$d(\%) = \frac{d \times 100}{M}$$

Relative error: The sum of the errors ΣE divided by N and the result by M,

$$E_{rel} = \frac{\Sigma E/N}{M}$$

Average deviation:

$$d_{av} = \frac{\Sigma d}{N}$$

This is numerically related to the standard deviation but has no real statistical significance.

Standard deviation σ:

$$\sigma = \sqrt{\frac{\Sigma(d^2)}{N - 1}}$$

[This may be expressed as the percent relative standard deviation (%RSD) = $(\sigma/M) \times 100$.]

Example 1.1 illustrates some of the statistical values obtained from the treatment of three observed values.

Example 1.1

V	Mean	Error	Deviation	Average deviation d
103	309/3	0	0	
101		−2	2	
105		+2	2	4/3
309	103	0	4	1.33

In Example 1.1, $E_{rel} = \Sigma (E/N)/M = (4/3)/103 = 4/309$; $d(\%) = [4/(3 \times 103)] \times 100\%$; $E = 0 - 2 + 2 = 0$. It can be seen that there are no determinate errors remaining because $E = 0$. The indeterminate errors, however, give rise to deviation of random differences from the true answer.

3. Standard Deviation

The most commonly used method of presenting the reliability of results is in terms of sigma (σ). In Fig. 1.7, the standard deviation coincides with the point of inflection of the curve; 68% of the results obtained fall within $\pm\sigma$ of the true answer, and 95% of the results obtained fall within $\pm 2\sigma$ of the true answer. These facts enable us to give some meaning to quantitative analytical results.

For example, let us suppose that we know by previous testing that the standard deviation of a given analytical procedure for the determination of silicon (Si) is 0.1%. Also, when we analyze a particular sample using this method, we obtain a result of 28.6% Si. We can now report with 68% confidence that the true analysis is 28.6 ± 0.1%. We could further report with 95% confidence that the true result is 28.6% ± 0.2% (where 0.2% = 2σ). Hence the report that an analysis of a sample indicated 28.6% Si and that 2σ for the method is 0.2% means that we are 95% confident that the true answer is 28.6% ± 0.2% Si. It is of interest to note that 99.7% of the results are within 3σ and 99.99% are within 4σ, i.e., only 1 result in 10,000 could be outside 4σ of the answer.

Such information concerning the reliability of the method is called *precision data*. Analytical results published without such data lose much of their meaning. They indicate only the result obtained, and not the reliability of the answer.

4. Precision and Accuracy

It is very important to recognize the difference between precision and accuracy. *Accuracy* is a measure of how close a determination is to the true answer. *Precision* is a measure of how close a set of results are to each other. The difference is illustrated in Table 1.14. A superficial examination of the results provided by analyst 2 could be misleading. It is easy to be deceived by the closeness of the answers into believing that the results are accurate. The closeness, however, shows that the results of analyst 2 are *precise*, and not that the analysis will result in our obtaining the true answer. The latter must be discovered by an independent method, such as having analyst 2 analyze a solution of known composition (e.g., a solution of pure chemical) and checking his answer against the known composition of the chemical. The U.S. National Bureau of Standards in Washington, D.C., has a number of such samples available for analysis.

It is important to realize that the inability to obtain the correct answer does not necessarily mean that the analyst uses poor laboratory technique or is a poor

Table 1.14   Results Obtained[a] (%)

	Analyst 1[b]	Analyst 2[c]	Analyst 3[d]
	10.0	8.1	13.0
	10.2	8.0	9.2
	10.0	8.3	10.3
	10.2	8.2	11.1
	10.1	8.0	13.1
	10.1	8.0	9.3
Average	10.1	8.1	11.9
Error	0.0	2.0	0.9

[a]True answer is 10.1% (obtained independently).
[b]Results are precise and accurate.
[c]Results are precise but inaccurate.
[d]Results are imprecise and inaccurate.

chemist. Many causes contribute to poor analyses, and only by being honest in recording the results can the causes of error be recognized and eliminated.

5. Error Analysis

When a new analytical procedure is developed or when an analytical procedure already in use is put into operation for the first time in a particular analytical laboratory, it is necessary to determine what errors are involved in the method. First, it is necessary to detect and eliminate any determinate error. The results obtained by the new method are then compared with those of an established method. If they agree, the new method is free from determinate error; if they do not agree, an error is involved in the new method.

Each of these methods works well when the error involved is significant and obvious. However, when the difference between the results is small, the following question arises: Is the error caused by a small determinate error or an indeterminate error? This difficulty can be resolved by carrying out a number of analyses by the new method. Using a statistical approach, we can determine if the error is significant and therefore involves a determinate error, or if it is merely part of the distribution of results encountered with indeterminate errors. If a determinate error is involved, it must be eliminated.

Having detected the presence of a determinate error, the next step is to find its source. Practical experience of the analytical method or first-hand observation of the laboratory operator using the procedure is invaluable. Much time can be wasted in an office guessing at the source of the trouble. Unexpected errors can be discovered only in the laboratory.

c. Common Determinate Errors

Operator Error. Operator errors are caused by the analyst performing the analysis. They may be the result of inexperience; that is, the operator may use the equipment incorrectly—for example, by placing the sample in the instrument incorrectly each time or setting the instrument to the wrong condition for analysis. Some operator-related errors are (1) *carelessness*, which is not as common as is generally believed, and (2) *poor sampling*, which is a common error. If the sample taken to the laboratory is not representative of the complete sample, no amount of good analysis can produce the correct result. Special attention should be paid to taking a representative of the complete sample for analysis. For example, it is usually fatal to send an untrained assistant to take a sample. It is equally dangerous to take the "first cupful" as a good sample. (3) *Sample handling* is also important. After a representative sample is received at the laboratory, it should be stored properly. Incorrect storage (e.g., in an open container) can cause contamination by impurities in the air or evaporation of the volatile components of the sample. After a while, the stored material is not representative of the original sample. Another storage problem may be temperature: A sample that is kept too warm may decompose. Remember that even good wine goes sour if it is stored incorrectly. Each of the foregoing errors can be eliminated by proper operator attention.

Reagents and Equipment. Determinate errors can be caused by contaminated or decomposed *reagents*. Impurities in the reagents may interfere with the method. The reagents may also be improperly labeled. The suspect reagent may be tested for purity by using a different set of reagents to recheck the work and by checking to see that the procedure used was the same as that recommended.

Numerous errors involving equipment are possible, including incorrect instrument alignment, incorrect wavelength settings, incorrect reading of values, and incorrect settings of the readout scale (i.e., zero signal should read zero, and for many samples the maximum signal should be set at 100 or some other suitable number). Any variation in these settings can lead to repeated errors. These problems can be removed by a systematic check of the equipment.

Analytical Method. The analytical method proposed may be unreliable. It is possible that the original author obtained good results by a compensation of errors; that is, although he appeared to obtain accurate results, his method may have involved errors that balanced each other out. When the method is used in another laboratory, the errors may differ and not compensate for each other. A net error in the procedure may result. Errors involved include (1) incomplete reaction for chemical methods, (2) unexpected interferences from the solvent or impurities, (3) a high contribution from the blank (the *blank* is the result obtained by going through the analytical procedure, adding all reagents, and taking all steps but either adding no sample or using pure solvent as the sample), (4) incorrect choice of wavelength for measurement of spectra, (5) an error in calculation based on incorrect assumptions in the procedure (errors can evolve from assignment of an incorrect formula or molecular weight to the sample), (6) matrix interferences, and (7) unexpected interferences. Regarding the last, most authors check all the compounds likely to be present to see if they interfere with the method; unlikely interferences may not have been checked. Checking the original publication should clear up this point.

There are other sources of error, such as the use of contaminated distilled water in a chemical method or a change in the line voltage used to operate the instruments. The latter is particularly likely in industrial areas where heavy demands on local power may be made, or released, suddenly. This problem may be acute when the work shift changes or when plants begin operating in the morning or close down for the night. At such times the switching on or off of heavy machinery may significantly change the line voltage available in the laboratory.

By careful checking, all of the foregoing sources of error can be detected and eliminated.

In an alternative method of calibration, the calibration curve produced by standard addition (Fig. 1.8) may be extended backward. The x intercept indicates the "concentration" in the sample, in this case, 2.23 ppm. A parallel line is drawn through the background contribution (0.5), indicating a contribution of 0.38. The corrected sample concentration is 2.23 − 0.38 = 1.85 mg/100 ml.

As yet another alternative, a second line parallel to the y axis and passing through the background measurement (0.5) is drawn. The concentration is read at the point of intersection between the line described and the extended calibration curve as shown in Fig. 1.8.

6. Errors Associated with Beer's Law Relationships: The Ringbom Plot

Whenever a measurement is made, there is always an associated indeterminate error. This is as true of measurements of radiation as any other measurement. These errors may involve the light source, the monochromator, the detector system, or other parts of the system, but, because they are indeterminate errors, they are not correctable. The errors in measurements of radiation intensity lead directly to errors in the measurement of the concentration when using calibration curves.

Let us suppose that the absolute error is a constant with respect to the radiation intensity falling on the detector and is equal to 1 unit of recording chart paper. At high concentrations of sample, the degree of absorption is high and very little radiant energy falls on the detector. If the amount of energy falling on the detector is equivalent to 2 units, then the error involved is 1 unit. This is a 50% relative error.

At the other end of the scale (see Fig. 1.9), if the concentration is very low, then the amount of radiation falling on the detector is high (e.g., 97 units), corresponding to a 3% absorption. If there is a 1-unit error in the measurement, then the radiation intensity falling on the detector may be 96 units, corresponding to a 4% absorption. The error in measuring the percent absorption is 33% and there is a corresponding relative error of 33% in determining the concentration in the sample from the calibration curve. These two examples occur at the extreme ends of the concentration ranges used in a Beer's law relationship, and it can be seen that significant error is involved.

We can plot this relative error as a function of transmittance, and it will give us a curve such as that shown in Fig. 1.10. This is known as a _Ringbom plot_. The minimum relative error occurs at 37% transmittance, although satisfactory results can be achieved over the range of 15–65% transmittance, that is, an absorbance range of 0.82–0.19.

As a consequence of this relationship, it is always advisable to determine the concentration of samples from absorbance readings that lie within this range. If the samples are too dilute, they may be concentrated by solvent extraction or evaporation. If they are too concentrated, they may be diluted to bring them into the

Table 1.15 Nomenclature and Definitions for Spectroscopy Experiments

Term	Symbol	Definition
Transmittance	I_1 I_0	Ratio of light intensity after passing through sample, I_1, to light intensity before passing through sample, I_0
Absorbance	A	$-\log T = abc$
Absorptivity	α	A/bc, where b = path length and c = concentration (g/liter)
Path length	b	Optical path length through sample
Sample concentration	c	Concentration of sample (g/liter)
Molar absorptivity	ξ	A/bc, where c = concentration (mol/liter)
Absorption maximum	λ_{max}	Wavelength at which highest absorption occurs
Wavelength	λ	Distance between wave crests
Frequency	ν	Number of waves per unit time
Wave number	$\bar{\upsilon}$	Number of waves per centimeter
Wavelength unit	$\mathring{A}$	Angstrom, 10^{-10} m
	nm	Nanometer, 10^{-9} m or 10 $\mathring{A}$
	μm	Micrometer, 10^{-6} m

desirable range. However, if it is not possible to alter the samples and they are outside the range indicated, the only alternative is to continue the analysis, knowing that the precision of the procedure will be diminished depending on the percent transmittance measured.

Recommended nomenclature and definitions for spectroscopy experiments are given in Table 1.15.

a. Errors and Relative Errors in Spectrophotometry

The actual readings made with a spectrophotometer are of the intensities of I_0 and I_1. The ratio of these two numbers gives the value of the transmittance. A typical transmittance–concentration curve is shown in Fig. 1.8. There are several sources of error in measuring I_1 (and therefore in measuring transmittance). One of the chief sources of error is the absorption of radiation by other compounds in the sample. Other sources of error include faulty readout from the detector, incorrect wavelength setting, variation in I_0 during measurement of I_1, and faulty sample preparation. In short, any measurement of transmittance on absorbance is subject to error.

Examination of Fig. 1.9 indicates the magnitude of the concentration error that results from a 1% error in measuring T. The smallest absolute error in transmittance occurs at the lowest concentration. The highest absolute error in concentration occurs where the curve is relatively flat. In this part of the curve, small

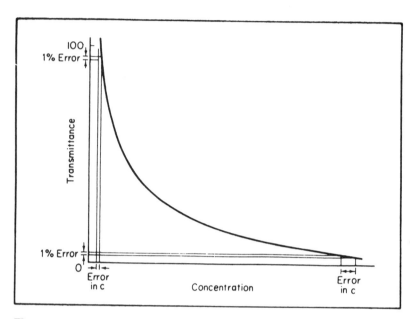

Figure 1.8 Relationship between the relative error in concentration (%) and the transmittance (Ringbom plot) with a constant error in T.

errors in measuring T result in large errors in measuring concentration. However, the relationship between an error in measuring T and the *relative error* in measuring the concentration of the sample is different. In this instance,

$$\text{relative error} = \frac{\text{error in measuring the concentration}}{\text{true value of the concentration}}$$

At the steep part of the curve in Fig. 1.9, 1% error in transmittance results in a low absolute error in concentration. However, the concentration is also very low. A small absolute error may therefore be large in relation to the concentration being measured.

The relationship first noted by Ringbom between the percent of relative error and the transmittance error can be calculated from Beer's law. The error in measuring the transmittance is termed the *photometric error*. Numerically, it can be shown that

$$\frac{\text{relative error (\%)}}{\text{photometric error (\%)}} = \frac{\Delta c / c3}{\Delta T}$$

where c is the concentration and T is transmittance; but from Beer's law

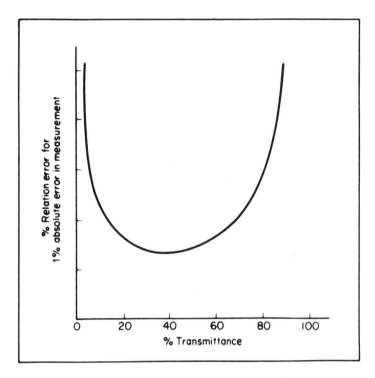

Figure 1.9   Ringbom plot relating relative error in concentration and the transmittance measurement with a constant 1% absolute error in T. The minimum relative error is at about 37% transmittance. It is acceptable between 15 and 65%.

$$A = abc \text{ or } ab = \frac{A}{c}$$

It can be shown that

$$\frac{\Delta c/c}{\Delta T} = \frac{1}{TA}$$

Using log base e,

$$-\ln T = A = abc$$
$$\ln T = -abc$$
$$\frac{\Delta T}{T} = -ab \, \Delta c$$

or

$$\Delta c = - \frac{\Delta T}{T} \frac{1}{ab}$$

But $A = abc$; therefore

$$\Delta c = \frac{\Delta T}{T} \frac{c}{A}$$

$$\frac{\Delta c}{c} = \frac{\Delta T}{TA}$$

$$\frac{\Delta c/c}{\Delta T} = \frac{1}{TA} = \frac{1}{T \ln T^{-1}}$$

$$= \frac{0.4343}{T \log T^{-1}} = \frac{0.4343}{T \log (I_0/I_1)}$$

The expression $\Delta c/c$ is the relative error in concentration resulting from an error ΔT in the measurement of transmittance. The relationship (shown in Fig. 1.8) discloses three important facts: (1) Relative error is very high at both very high and very low values of T (this occurs when the sample concentration is very low or very high); (2) the relative error is lowest over the 20–60% transmittance range; and (3) the relative error is a minimum at 36% transmittance (absorbance = 0.434).

For best results, it is advisable to work in the 20–60% transmittance range. If the sample is too concentrated (T too low), simple dilution may be effective. If the sample is too dilute (T too high), an extraction, evaporation, or other means of concentrating the sample may be required. Care should be taken not to introduce errors into the analytical method while performing such concentration steps. If none of these steps are possible, the results obtained should be reported with a note indicating the loss of reliability of the data under these conditions.

b. Detection of Indeterminate Error

Indeterminate errors are always present. The extent of the errors is determined by statistics as described earlier. The standard deviation can be determined accurately only by carrying out an infinite number of tests, but this is not a practical proposition. However, an *estimate* of σ can be made from a finite number of observations by using the equation

$$\sigma = \sqrt{\frac{\Sigma(d^2)}{N - 1}}$$

This "estimate of σ" is always greater than "true σ" to allow for the uncertainty of *estimation* of σ from a finite number of tests compared to the *determination* of σs from an infinite number of tests. An increase in the number of observations brings about a decrease in the error in the estimate, so that the calculated value approaches true σ.

A reduction in σ can somtimes be brought about inadvertently when eliminating determinate error. A skillful operator might improve his skill and reduce operator-determinate error to a small value and simultaneously reduce operator-indeterminate error. This is why one operator can carry out an analysis using a particular method and obtain a low standard deviation, whereas another, using the same method and sample, might get results with a higher standard deviation.

In practice, the standard deviation for a particular method used for a particular sample may be low when the determination is carried out by a single operator or by the author of the method. If a method is used by 10 different operators over an extended period of time, however, the standard deviation will be significantly higher, even if the same sample is used each time. These two sets of precision data are called *short-term precision* (obtained by one operator at one time) and *long-term precision* (obtained by several operators at several different times, or even one operator over a long period of time).

Precision can be improved by paying closer attention to the details of the procedure, operating the equipment at peak performance conditions, using correct sampling procedures, and storing the sample properly.

c. Rejection of Results

Another source of imprecision is the occasional result that is obviously in error. This result may have been caused by incorrect weighing or measuring, spillage, faulty calculation of results, or carelessness. In any case, the result that includes such a gross error should be rejected and not used in any compilation of results. An acceptable rule to follow is that if the error is greater than 4σ, it should be rejected. When calculating the standard deviation for a new procedure, a suspected result should be included in the calculation of σ. After the calculation, the suspected result should be examined to see if it is more than 4σ from the true value. If it is outside this limit, it should be ignored; if it is within this limit, the value for σ should be recalculated with this result included in the calculation. It is not permissible to reject several results on this basis.

d. Significant Figures

Analytical results, among many other things, are reported in numbers, such as 50.1%, 10 parts per million (ppm), or 25 ml/liter. The numbers should be such that there is uncertainty only in the last figure of the number. For example, the number 50.1% means that the percentage is closer to 50.1 than to 50.2 or 50.0, but it does not mean that the percentage is exactly 50.1. In short, we are sure of the "50" part of the number, but there is some doubt about the last figure. If we were to analyze two samples containing 50.08% and 50.12% of a component by using an instrument accurate to 0.1%, we would not be able to distinguish the difference in the compositions of the samples, but would report them both as 50.1%.

The number 50.1 has three significant figures (5, 0, 1). Since there is some doubt about the last significant figure, there is no point in reporting any more figures (even though they might be mathematically obtainable), because they would be meaningless. For example, in the sample discussed above, experimentally there was shown to be 50.1% of component A. It was known from other sources that component B was 25% of component A. There should therefore be (50.1 × 25)%/100 of component B, or 12.525%. The number 12.525 was derived from a number with three significant figures, the last of which was in the tenths of a percent. The calculated value cannot contain more than three significant numbers; therefore we should report that component B was present at 12.5%.

The reporting of figures implies that all the numbers are significant and only the last number is in doubt, even if that number is zero. For example, $1.21 × 10^6$ implies that the number is closer to $1.21 × 10^6$ than to $1.22 × 10^6$ or $1.20 × 10^6$. But 1,210,000 implies that the number is closer to 1,210,000 than to 1,210,001 or 1,209,999. Furthermore, the number 50.10 implies 10 times greater accuracy than 50.1.

The following are some rules that should be observed when reporting results.

1. In enumerating data, report all signficant figures, such that only the last figure is uncertain.
2. Reject all other figures, rounding off in the process. That is, if a number such as 1.325178 must be reported to four significant figures, the first five figures should be considered. If the fifth figure is greater than 5, increase the fourth figure by 1 and drop the fifth figure. If the fifth figure is less than 5, the fourth figure remains unchanged and the fifth figure is dropped. If the fifth figure is 5 and the fourth figure is even, it is not increased when the 5 is dropped. Table 1.16 shows some examples of rounding to four significant numbers.
3. In reporting results obtained by addition and subtraction, the figures in each

Table 1.16 Rounding Off to Four Significant Figures

Number	Four significant figures
1.37285	1.373
1.27245	1.372
1.3735	1.374
1.3725	1.372
1.37251[a]	1.373

[a]The number 0.00051 is greater than 0.0005, even though the last figure is not significant; hence the fourth figure is increased by one.

number are significant only as far as the first doubtful figure of any one of the numbers to be added or subtracted. For example, the numbers in the set 21.1, 3.216, and 0.052 are reliable to the first decimal point; that is, 21.1 is reported as 21.1, 3.216 is rounded off to 3.2, and 0.052 to 0.1. The sum of the rounded-off numbers (21.1 + 3.2 + 0.1) is 24.4. In the first number, uncertainty arises at the tenths place. All other numbers are rounded off to the tenths place prior to addition.

4. For multiplication and division, the number of significant figures in each term and in the answer should be no greater than that of the term with the least number of significant figures. For example, $1.236 \times 3.1 \times 0.18721 \times 2.36$ is rounded off to $1.2 \times 3.1 \times 0.2 \times 2.4$. The answer is rounded off to 1.7. In this case, the term 3.1 contains only two signficant figures. It is meaningless to give the other terms or the answer to more than two significant figures.

5. The characteristics of a *logarithm* indicates an order of magnitude; it is not a significant number. The mantissa should contain no more significant numbers than are in the original number. For example, the number 12.7 has a log of 1.1038. The log is rounded off to three significant figures and is reported as 1.104.

6. If several analyses have been obtained for a particular sample (*replicate analysis*), it should be noted at what point there is doubt in the significant numbers of the result. The final answer should be reported accordingly. For example, given the triplicate answer 11.32, 11.35, 11.32, there was no doubt about 11.3, but there was doubt about the fourth figure. The average should be reported as 11.33 [i.e., $(11.32 + 11.35 + 11.32) \div 3$]. Given the triplicate answer 11.42, 11.35, 11.22, there was no doubt about 11, but there was doubt about the next figure. The average should be reported as 11.3 [i.e., $(11.4 + 11.3 + 11.2) \div 3$].

H. SIGNAL AND NOISE (SENSITIVITY)

If a recording is made of a signal versus wavelength or a signal versus time, the ideal tracing would be a smooth line and a peak, as shown in Fig. 1.10. In practice, however, the recorded trace is seldom smooth, but includes random signals called *noise*. Noise can originate from small fluctuations in the light source, the detector, stray light, the recorder, the monochromator system, and so on. Provided that the signal is significantly greater than the noise, it is not difficult to make a reliable measurement of it.

In Fig. 1.10c the noise level is increased and is superimposed on the signal. The value of the signal is less certain. The results are less precise, but clearly discernible. In Fig. 1.10b the noise level is greater than the signal, and it is virtually impossible to make a meaningful measurement of the latter.

The noise can be reduced by optimizing the equipment settings (i.e., source,

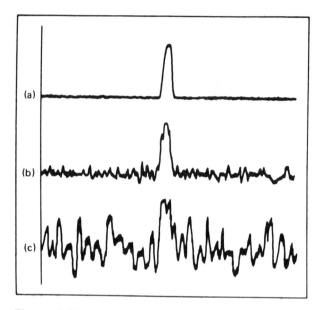

Figure 1.10 Signal versus wavelength with different noise levels: (a) no detectable noise, (b) moderate noise, and (c) high noise.

monochromator slit, detector, amplifier, etc.). It may further be improved by "smoothing" out the signal with a condenser. Unfortunately, this has the effect of slowing the response of the instrument. So a trade-off is usually made by the operator, who uses as much smoothings as possible but preserves a response time fast enough to be useful.

The sensitivity of an analytical procedure is ultimately limited by the noise level. When the signal can no longer be distinguished from the noise, the sensitivity limit of the system has been reached.

It is conventional to state the signal-to-noise ratio when defining the sensitivity of a method. This is an arbitrary decision that varies from one worker to another. Generally, a ratio of 1:1 is used (i.e., signal = noise), but other ratios are also commonly used.

A method that has improved the sensitivity limit is *time averaging*. In this case, a signal, such as that illustrated by Fig. 1.10c, is recorded 100 or more times very rapidly and overlapped and stored in a computer. Since the noise level is random, it averages out to a curve like Fig. 1.10a. The signal itself, however, is not random, but additive. Each run results in an accumulated signal that increases with the number of scans. Using this technique, signals that would normally be lost in the noise can be detected and measured. The technique has been used in NMR, UV, IR, and nuclear science methods.

1. Signal-to-Noise Ratio

The signal-to-noise ratio (S/N) is a limiting factor in making absorption or emission measurements. When the signal is not discernible from the noise level, it is impossible to make any meaningful measurements. In the Fourier transform system, there is an immediate advantage in the fact that the signal intensity falling on the detector is high. The detector operates under optimum conditions and therefore there is a reduction in noise level.

Furthermore, the S/N level is proportional to the square root of T, where T is the time over which signal and noise are accumulated. If an observation time of 1 sec is necessary to give an S/N of 1:1, this must be increased to 4 sec to give an S/N of 2:1. It can be seen that there is some advantage to this system, but logistics prevent very long time measurements from being made.

A second way to improve S/N is to use repetitive scans. In this instance advantage is again taken of the fact that noise is random but the signal is additive. With the Fourier transform system, a complete scan can be taken using a single burst of radiation from the light source, permitting rapid scanning and recording on computer tape over a reasonable time period after a laser source is used. If a scan is taken many, many times and each instance is recorded on the same piece of computer tape, then the signal will increase, whereas the noise level will tend to average out. If the scans are of a short enough duration, it is possible to take hundreds of scans in a relatively short time. The relationship between S/N and the number of scans is shown in Fig. 1.11.

A third method to improve the signal to noise level is to add a condenser between the detector output and the amplifier. The net effect is to average out the noise level, but the signal accumulates. The capacity of the condenser used depends on the required reduction of the noise level. The greater the capacity, the better the smoothing of the signal. Unfortunately, the system is limited, because as the capacity of the condenser increases, the response time of the system also increases. Transient signals are often not detected at all. In practice, the operator must select a compromise capacity so that the optimum response time and S/N are obtained.

The decision to improve S/N is always an arbitrary one. It is theoretically possible to measure any signal in any noise level, provided that the measurements are taken over a long enough period of time and repeated enough times to accumulate the signal. In practice, there is usually a time limitation on such observations and the operator must make this decision based on the circumstances under which he is working.

The Fourier transform system has found many applications, not only in IR spec-

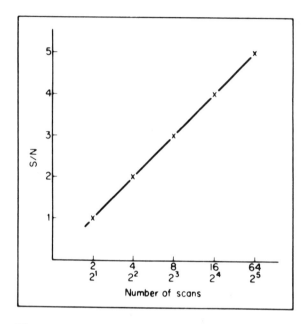

Figure 1.11 The signal-to-noise ratio increases with the square of the number of observations or scans.

troscopy, but also in NMR, UV, and so on. The disadvantage of this system is that the readout from the detector does not in any way resemble an IR spectrum. It is composed primarily of an infinite number of interference patterns and it is necessary to use a computer to unscramble these in order to reveal the true IR spectrum. Programs are available to do this, however, so it is no longer a practical problem.

I. SAMPLING

The most important single step in analysis is obtaining a representative sample of the material to be analyzed. If a nonrepresentative sample is taken, no matter how excellent the analytical procedure may be or how expert the analyst, the result obtained will be incorrect. There are two requirements for correct sampling. First, a sample must be taken that is sufficient for all the analyses to be carried out comfortably with enough for at least duplicate analyses if necessary. Of course, if only a small quantity of sample is available, the analyst must do his best with what is provided. Second, the sample must be representative of the material being sampled. If the sample does not represent the material, the analysis cannot provide a reliable characterization of the material and the result is in error regardless of the accuracy of the analytical method.

1. Sampling of Different Sample Types

a. Gas Samples

Two types of gas samples can be taken. First, single spot samples can be taken in a balloon or syringe after the sample has been well stirred with a fan or paddle. Several composite samples of gas can be taken by bleeding gas slowly into a suitable container and collecting the gas over a period of time. If necessary, this can be done over several hours and an average analysis of the gas stream obtained.

b. Liquid Samples

It is important to stir liquid samples adequately and to take samples remote from sources of contamination. For example, a liquid brought to the laboratory in a container should be sampled so as to avoid floating froth or sludge on the bottom of the container. Spot or composite samples can be taken in a manner similar to that used for gas samples. River or sea samples should be taken at points where contamination is likely to be minimal. With river samples it is important to avoid river bank contamination, floating froth contamination, or concentrated discharge from manufacturing processes. River and sea water samples should be taken at several depths and distances from the shore. Continuous analysis of commercial products on stream should be taken by placing the analysis probe in a position where the sample is representative of the product.

c. Solid Samples

Solid samples are the most difficult to handle because they cannot be conveniently "stirred up." Moreover, unlike the situation with fluids, there are no diffusion or convection currents in solids to ensure some mixing. Numerous methods have been devised to reduce this problem. The recommended method should be used for each particular type of sample, such as polymers, metals, soils, cements, foods, cloth, plant materials, and biological specimens. Frequently, this involves taking portions from different parts of the sample, mixing them, and making a representative part.

d. Bulk Samples

The process of sampling bulk materials, such as coal, metal ore, and soil samples, requires several steps. First, a gross representative sample is gathered. The size of the representative sample should be at least 500 times as big as the largest particle in the bulk sample. Numerous portions of the sample should be taken from various locations. The combination of these portions should be representative of the whole sample. With soil samples from a field, drillings are taken from different areas of the field. Unique areas should be represented but not overemphasized.

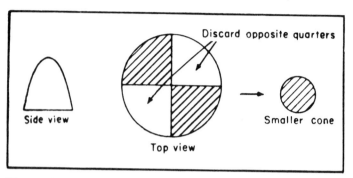

Figure 1.12 The cone and quarter method of sampling bulk materials.

The gross representative sample may be too large to transport or to handle in the laboratory, so it must be reduced in size. Large particles (as in coal) should be broken up and mixed into the main sample. A smaller representative sample may be obtained by several methods. In the *long pile and alternate shovel method*, the sample is formed into a long pile. It is then separated into two piles by shoveling alternate shovelfuls into first one pile and then the other. One pile is discarded. If necessary, the second pile may be further reduced by additional piling. The *cone and quarter method* is widely used. In this method, the sample is made into a circular pile and mixed well. It is then separated into imaginary quadrants. A second pile is made up from two opposite quarters of the first, the remainder of which is discarded. This process is shown in Fig. 1.12. As the total size of the sample is reduced, it should be broken down to successively smaller pieces. The final sample should be representative of the sample and large enough to provide sufficient material for all the necessary analyses.

Metal samples need special care. When a molten metal solidifies in a crucible or other container, the first solid metal formed tends to be purer than the remainder. The last metal to solidify is the most impure and is generally located in the center, or *core*. It is important to bear this in mind when sampling casting or pressings. If convenient, a sample should be ground from a representative cross section of the sample. Otherwise, a hole may be drilled across the sample at a suitable place and the drillings mixed and used as the sample.

2. Storage of Samples

When samples cannot be analyzed immediately, they must be stored. The storage container must be clean and airtight. Liquid or gas samples can be stored in plastic containers. Corrections must be made for any absorption effect on the walls of the container, particularly in trace metal analysis of fluids. Solid samples should also be sealed. All samples should be stored in rooms that are not too hot or too humid. Storage for long periods of time should be avoided, because decomposition or evaporation of the sample may take place. Special care should be taken with samples containing trace quantities (parts per million) of materials that must be determined subsequently. Trace components may plate out on the glass or plastic vessel upon standing or may be contaminated by leaching impurities from the wall of the container. The problem can be overcome by freezing the samples.

REFERENCES

1. Newton, I., *Optics or Retreaties of the Refractions, Inflections and Colors of Light,* 1704. Revised by Dover Publications, New York, 1952.
2. Wollaston, W. H., *Phil. Trans. of The Royal Society,* London Series A. 92:365 (1802).
3. Walsh, A., *Spectro Chim Acta* (1953).
4. Chandler, C., *Atomic Spectra,* D. Van Nostrand, New Jersey, 1964, Appendix 4.

2
Atomic Absorption Spectroscopy

Atomic absorption spectroscopy involves the study and measurement of the absorption of radiant energy by free atoms. The data obtained by studying this absorption provide spectroscopic and analytical information. The spectroscopic information includes the measurement of atomic energy levels, the determination of oscillator strengths, the population of atoms in various energy levels, atomic lifetimes, and so on. The analytical information revealed includes qualitative and quantitative determination of elements, particularly the metallic elements of the Periodic Table.

It is the objective of this chapter to present atomic absorption spectroscopy principally as an analytical tool. To do this, it is necessary to understand both the analytical chemistry and the spectroscopy involved in order to generate reproducible reliable analytical data.

The analytical process involves the conversion of molecules or ions into free atoms and the measurement of absorption of radiation by these free atoms. The conversion of solutions of chemical compounds to free atoms involves areas of physics, chemistry, and astrology—the latter still playing the most important role even after more than three decades of study.

As indicated earlier, Walsh made major contributions in converting a physical phenomenon, first observed by Foucault in 1849, into an analytical technique by introducing hollow cathode light sources, flame atomizers, photomultipliers, and a modulated system. He demonstrated that the system worked as an quantitative analytical tool(1).

Atomic absorption using flame atomizers was rapidly accepted as a simple technique which gave high sensitivity, accuracy, and precision and was deceptively easy to use even in the hands of modestly trained operators.

In 1961 Lvov published details of the first carbon atomization system. His instrument showed a sensitivity improvement of several orders of magnitude over flame atomizers. But good quantitative data were difficult to obtain. In 1969 several authors published details of carbon atomizers at an International Conference on Atomic Spectroscopy in Sheffield, England.

The instrument manufacturers, glowing with the lucrative success of the flame atomizer, rapidly developed and marketed commercial carbon atomizers. Unfortunately, the difficulties in controlling carbon atomization were greatly underestimated, and to this day the commercial models suffer from the problem of needing a high degree of skill to obtain reliable, quantitative data. As Herb Kahn says, "the difference between using flame atomizers and carbon atomizers is the same difference as between watching tennis and playing tennis." However, it still remains perhaps the most sensitive analytical technique for the quantitative analysis of metals commercially available today.

A. GENERAL PRINCIPLES

The principles of atomic absorption spectroscopy have been discussed in Chapter 1. We should remember that the process involved generating free atoms using an atomizer, which may be flame or a thermal atomizer. Radiation from a suitable light sources such as a hollow cathode lamp is then shone through the atomic population produced and the degree of absorption of the radiation is measured. The absorption is measured at one of several wavelengths characteristic of the element being determined. Absorption by free atoms at other wavelengths is essentially zero. The wavelength is selected using a monochromator and is measured using a photomultiplier.

B. EQUIPMENT USED FOR ATOMIC ABSORPTION

1. The Optical System

Atomic absorption spectroscopy uses the same basic optical system as other forms of absorption spectroscopy. The components include a radiation source (a sample "container"), a monochromator to select the desired wavelength of radiation and limit the spectral slit width, and a detector for measuring the intensity of radiation after it passes through, and is absorbed by, the sample. The components can be lined up in two basic optical systems: (1) single-beam optics and (2) double-beam optics.

a. The Single-Beam Optical System

A single-beam optical system was shown in Fig. 1.3. Radiation from the light source passes through the sample atoms generated in the atomizer. From here the light proceeds to a monochromator which consists of a dispersion element and a slit system. The dispersion element—e.g., a prism on grating— separates the radiation by wavelength. Using a slit system, only light of the desired wavelength is allowed to proceed along the light path of the detector. The detector in turn measures the intensity of the light, and the signal is read out on a recorder.

A typical absorption signal obtained from a zinc sample is shown in Fig. 2.1. In a typical experiment it may be seen that with no zinc present the light intensity is approximately 100 units. When zinc is introduced into the system, the signal intensity drops to approximately 95 units, showing a loss of 5 units of intensity. This is 5% absorption; that is, 5% of the light that entered the system was absorbed by the atoms.

The single-beam system works well in practice and is commonly used. However, it is subject to one major source of error "drift" (see page 29). If the intensity of radiation from the source varies, then the entire signal drifts, as illustrated in Fig. 2.2.

This causes a significant change in the intensity of radiation reaching the detector. Also, drift in the response of the detector causes a change in the signal generated even for a constant light intensity. If the entire signal is being recorded, then it is not difficult to observe and correct for this drift; but if the percent absorption is read out on a digital readout system or a simple needle, then an error results. If the unabsorbed signal were 100 units, 5% absorption would give a signal of 95, but if the unabsorbed signal drifts to 93 and there is a 5% absorption by the sample, the reading would be approximately 88. An unsuspecting operator could

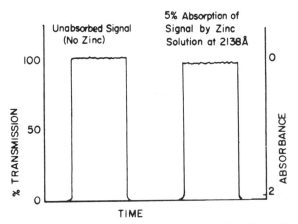

Figure 2.1 Absorption of 5% of signal by zinc solution.

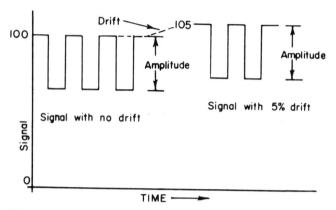

Figure 2.2   Change of signal with baseline drift.

mistake this for 12% absorption, since he would be under the impression that the original intensity was still 100. This source of error, which is due to "drift", can be largely overcome by using the double-beam system.

b. The Double-Beam Optical System

A schematic diagram of the double-beam system was shown in Fig. 1.4. In this system the light from the radiation source is split by a beam splitter and forms two paths: the reference path and the sample path. The beam splitting alternately directs the light from the radiation source along the reference path and then along the sample path. The light beams passing down these two paths are equal in intensity and pass alternately through sample and reference. After passing through the reference and sample they are then recombined and continue through a mono-chromator system to the detector.

In practice, the reference beam either is not absorbed (e.g., by using an empty reference beam) or is absorbed by a constant quantity by a standard reference cell. In atomic absorption, however, an empty reference beam is used. The radiation in the sample beam, however, is absorbed by the atomized sample. When the two beams are recombined, an oscillating signal is produced which falls on the detector.

If there is no absorption by the sample or by the reference, then the two beams recombine to form the original unsplit beam. This beam does not vary in intensity as the detector views alternately the reference beam and the sample beam. In this instance, the readout gives zero absorption.

When there is absorption in the sample beam, then the signal reaching the detector varies accordingly. Typical examples are shown in Fig. 2.3.

The amplitude is the difference in signal intensity between the reference and the

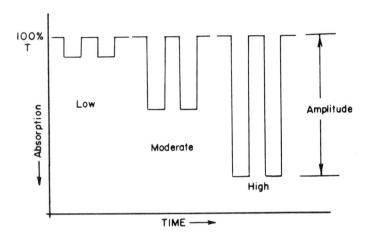

Figure 2.3 Typical signal reaching the detector with low, medium, and high absorption.

sample beam. It can be seen that the amplitude of the signal falling on the detector is greatly increased as the degree of absorption is increased.

The amplitude does not change appreciably when the intensity radiation from the source changes modestly. Let us take for example a sample that absorbs 10%, if the source intensity drifts 5%, but the differences between them is still 10% and there is no change in signal.

The great advantage of this system is that drifts do not cause a signal error in the readout. This can be seen in Fig. 2.2, which illustrates a signal with no drift in the radiation source and a 5% drift in the radiation source.

One of the difficulties with atomic absorption spectroscopy is that a suitable reference is not available for use in a double-beam system. All atomizers are dynamic in operation, and it is very difficult in practice to make an atomizer that will give a faithful reproduction of the background signal found in the sample atomizer. In practice, some commercial equipment still uses the single-beam system and relies on stabilized radiation sources and repeated checks in order to eliminate errors caused by drift. Other systems use a double-beam system that includes reference beam, but not a reference cell. In other words, the reference beam is empty, and in this case we are really using a pseudo-double-beam system.

Each optical system has worked well in operation and has been used to obtain reliable data. The single-beam system requires controlling the intensity of radiation from the hollow cathode over extended periods of time. This can best be done by using a highly controllable power supply to the hollow cathode system, which is frequently quite expensive. In contrast, the double-beam system, therefore, enables one to use less expensive equipment for the hollow cathode and power supply unit. However, one must also use the extra optical equipment necessary to incorporate a double-beam system. We are, therefore, faced with a common dilemma in spectroscopy: whichever way you go, some advantages are gained and some advantages are lost. The choice is left up to the purchaser.

2. Radiation Source

a. Absorption Line Width

The natural spectral width of atomic absorption lines is about 10^{-5} nm. This finite width reflects the natural energy spread of the ground state and the excited state in the atoms of interest under ideal conditions. The natural line width is broadened by the Doppler effect, which causes an increase in the absorption line width with increasing temperature of the atoms. It is common for absorption lines to be increased to about 0.001 nm by this effect. Collision between atoms causes a broadening of the energy levels in the ground state and the excited state and, hence, a broadening of the absorption bands. This is called *pressure broadening* and can be caused by collisions between like atoms (Holtzmark broadening) or different atoms (Lorenz broadening). It is common for pressure broadening to increase the line width to 0.002 nm. Absorption line widths are also dependent on the wavelength of the resonance line or absorption line. If the resonance line is at a long wavelength, the absorption line widths are greater than if they are at short wavelengths. This is illustrated in Table 2.1, where the absorption line widths for sodium and potassium are shown, representing short and long wavelengths and temperatures of 1000 and 3000 K.

In addition, local electrical fields cause broadening (Stark effect), and local magnetic fields also cause broadening (Zeeman effect). With proper shielding, these effects should be minor.

If a continuous source such as a hydrogen lamp or a tungsten lamp were used for atomic absorption, only the energy in the narrow absorption wave band

Table 2.1 Spectral Widths of Absorption Lines[a]

Element	Wavelength	Absorption line width	
		1000 K	3000 K
Na	589.5 nm	0.0028 nm	0.0048 nm
Zn	213.8 nm	0.0006 nm	0.001 nm

[a]Fundamental line width broadened by temperature and pressure.

indicated above would be absorbed. With a normal commercial instrument, a spectral wave band of about 0.1 nm falls on the detector. Normally, atoms absorb over a range of about 0.002 nm. If all of the light in this 0.002 nm absorption band width were absorbed from an emission wave band of 0.1 nm, the detector would record the loss of only 2% of the signal falling in it. The rest of the signal (0.098 nm) would not be absorbed by the free atoms because it is not included in the absorption line width. The sensitivity of an atomic absorption method is defined at the concentration that will give 1% absorption of the resonance line (width 0.002 nm). In this instance, using a continuous light source, the total energy falling on the detector would diminish only 2.0% and the detector would register a decrease in signal of the same magnitude. Since 100% absorption of the 0.002 nm absorption band results in a net absorption of 2% of signal, a 1% absorption of the 0.002 nm absorption band would result in a net absorption of 0.02%. Conventional detectors are not capable of measuring such small changes in signal with any reasonable degree of precision. Therefore, it is not possible to achieve either high sensitivity or useful quantitative data if a continuous source is used, particularly at shorter wavelengths.

This difficulty was overcome by Walsh (1), who demonstrated that hollow cathodes emit very narrow spectral lines. If the hollow cathode is made of the same element as the metal being determined, the line width of the hollow cathode radiation is somewhat narrower than the atomic absorption line width and, therefore, completely available for absorption. In addition, there is usually virtually no background light falling on the detector and the net signal approaches zero energy when the spectral line width is completely absorbed.

In summary, it can be said that atomic absorption line widths are very narrow and cannot be easily measured with conventional equipment using a continuous light source. A hollow cathode is necessary for operation although the demonstrated use of continuum light sources persists (2).

b. The Hollow Cathode

A schematic diagram of a hollow cathode, based on the design of Jones and Walsh (3), is shown in Fig. 2.4. In practice, the voltage is applied between the anode and the cathode, the anode being positive and the cathode negative. The system is filled with a noble gas, such as helium, argon, or neon. The filler gas, e.g., argon, is ionized by the anode and becomes a positive argon ion by the mechanism $Ar + e^- \rightarrow Ar^+ + e^- + e^-$. The positive ion Ar^+ is attracted by the negative cathode and accelerated under the influence of its charge. When it reaches the cathode it impinges on the metal surface, dislodging or "sputtering" excited metal atoms into the space inside the cathode. The excited metal atoms emit radiation at their own characteristic wavelengths and return to the ground state. The emitted radiation is used as the light source for the atomic absorption system. After the atoms return to the ground state, they form a cloud of free atoms and diffuse either back to the

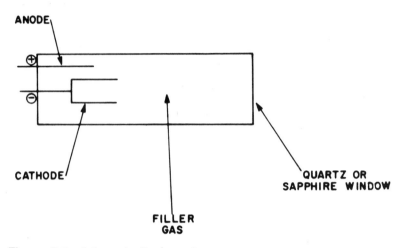

Figure 2.4   Schematic diagram of a sealed hollow cathode.

walls of the cathode or to the glass walls forming the hollow cathode envelope. If the atoms in the cathode are hot, spectral line broadening occurs caused by the Doppler effect. This is detrimental to the successful operation of the instrument, since the "wings" of these broadened emission lines cannot be absorbed by the free atoms from the sample.

i. Self-Absorption   A problem arising with the cold atoms inside the hollow cathode is that they are able to reabsorb some of the radiation emitted by the cathode itself. This absorption is at the very center of the emission line and is most easily absorbed. It leads to distorted calibrated curves because only the less easily absorbed parts of the lines emitted by the hollow cathode remain to be absorbed by the free atoms of the sample causing a relative decrease in % absorption by the same sample concentration. As will be described later in this section, this problem can be somewhat alleviated by using a demountable hollow cathode whereby the atom cloud is continuously removed by pumping.

An illustration of the effect of absorption by the atom cloud on the signal emitted from a hollow cathode is shown in Fig. 2.5.

ii. Intensity of Radiation Emitted by Hollow Cathodes   For a single-element hollow cathode, the intensity of the emitted radiation depends on the effectiveness of the bombardment of the hollow cathode by the charged filler gas to cause sputtering. The kinetic energy of an ion of bombarding filler gas must be greater than the energy necessary to dislodge the metal from the cathode surface, that is, the lattice energy of the surface metals. The lattice energy of the metal is a physical property of the metal and cannot easily be varied except by temperature. But, as

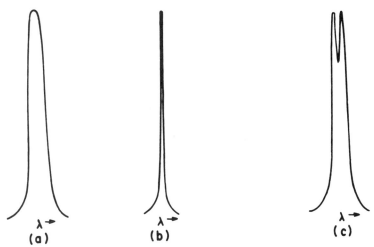

Figure 2.5 Distortion of spectral line shape in a hollow cathode. (a) Shape of spectral line emitted by a hollow cathode. (b) Shape of spectral energy band absorbed by cool atoms in a hollow cathode. (c) Shape of net signal emerging from a hollow cathode.

we have already seen, it is better to operate the hollow cathodes at low temperatures to avoid Doppler broadening.

The kinetic energy of the bombarding ion is directly controlled by the mass of the ion, the voltage across the electrodes, and the number of collisions per unit time that the ion experiences on its way to the cathode. The mean free path of the ion determines the duration of acceleration, and therefore its mean kinetic energy on reaching the cathode surface.

In practice, it is important that the geometry of the system, the mean free path of the ion (and therefore its pressure), and the voltage across the anode and cathode be carefully controlled in order to obtain a steady signal. The optimum voltages and pressures have been determined by careful research. It has also been found that the best shape for the hollow cathode is a tube sealed at one end and open at the other. The tube shape tends to keep free atoms inside the cathode, where they redeposit and therefore extend the lifetime of the hollow cathodde system. In addition, it provides an extended area over which atoms are emitting, and therefore increases the radiation intensity of the hollow cathode system.

iii. Filler Gas The most common filler gas used is argon, because as a rare gas it is readily available, the least expensive, and because it has a fairly simple spectrum. In addition, argon does not easily "harden" inside the hollow cathode, and therefore provides a system with a reasonably long life. Since argon emits its own radiation at characteristic wavelengths, there are occasions when an argon emission line coincides with, or is very close to, the resonance absorption line of the element being examined. This is so with lead hollow cathodes, for which helium is therefore recommended rather than argon, in order to avoid overlap of emission lines. Neon has been used in some cases, but the added expense usually precludes its use. Leakage of air into the system allows oxygen or nitrogen to become ionized rather than the filler gas. These ions are reactive, and they result in a decrease in sputtering. Ultimately this prevents operation of the system and must be avoided.

iv. Loss of Metal from the Cathode   The metal used in the hollow cathode is the source of radiation of the resonance line. With time this metal will all "sputter" away, leaving only the support material, which does not emit at the desired wavelength. The sputtering can be slowed by operating at low voltage. This is particularly important when volatile metals such as lead, tin, or arsenic are used. Commercial manufacturers have spent significant research time on this problem and now recommend conditions whereby a suitable emission intensity can be obtained while prolonging the lifetime of the hollow cathode system. It can be readily appreciated that increasing the voltage of the hollow cathode will increase its intensity for a short time but will also decrease its lifetime.

v. Hardening   The free atoms from the hollow cathode diffuse out into the open system and slowly trap the filler gas between themselves and the glass envelope. When this occurs the pressure inside the cathode decreases until it becomes too low for the system to work. Under these conditions it is said that the hollow cathode has "hardened." If the trapped filler gas can be released or if more gas can be introduced into the hollow cathode system, then the hollow cathode can be rejuvenated. With sealed systems, however, this is seldom possible.

vi. Demountable Hollow Cathode   One of the difficulties with the sealed hollow cathode lamp is that generally it is useful only for work with the element from which the cathode itself is made. Hence, if it is necessary to determine 10 different elements, then 10 different hollow cathodes are necessary. This is an expensive and time-consuming project, particularly in a research-oriented lab, where a number of different elements must be determined in the course of a day. The problem can be greatly alleviated by using the demountable hollow cathode lamp based on a design by Barnes, illustrated in Fig. 2.6. The advantage of the system is that the cathode itself can be easily removed and replaced by a cathode of another element in a few minutes. The system can be pumped down and operating again within 15–20 min. The filler gas flows constantly under controlled pressure through the system. Atom clouds are not allowed to build up, and hence self-absorption is greatly decreased. The initial outlay to buy the demountable hollow cathode is comparatively high, but since it is so easy and inexpensive to replace a cathode, the cost decreases over an extended time period. In addition, it

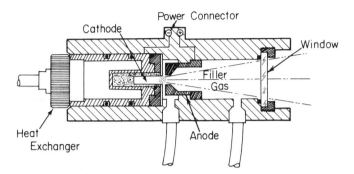

Figure 2.6 Barnes demountable hollow cathode lamp.

is very convenient to be able to change cathodes or filler gas with little difficulty and in a short time.

A problem encountered with volatile elements, such as Hg, Se, Pb, and As, was self-reversal resulting in loss of sensitivity. The problem was alleviated by drilling a hole in the back of the cathode and drawing filler gas through it. This reduced build-up of atoms and reduced self-absorption, enabling its use even for mercury—the most volatile metal (4). A design is shown in Fig. 2.7.

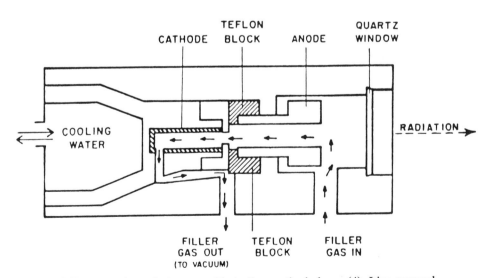

Figure 2.7 Flow-through demountable hollow cathode lamp (4). Line reversal is significantly reduced and more linear calibration curves obtained.

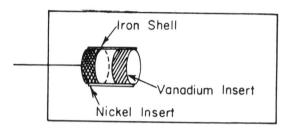

Figure 2.8 Multielement hollow cathode built from iron, nickel, and vanadium (5).

vii. Multielement Hollow Cathodes   It is frequently desirable to analyze two or three elements at the same time. This is a difficult procedure using single-element hollow cathodes. The problem can be overcome by using a multielement hollow cathode. The first such hollow cathode was made from iron, nickel, and vanadium (5) (Fig. 2.8). In practice, it was shown that emission spectra from all three elements were indeed generated and the resonance lines could be used for simultaneous analysis. However, it was found that inevitably one element was more volatile than the other two elements. With time, this element tended to cover over the surface originally occupied by the other two elements. Consequently the intensity of radiation from the other two elements decreased. The lifetime of the multielement hollow cathode was thereby greatly reduced. In addition, the steady drift in intensity presented analytical problems. At least some compatibility in the volatility of the metals is necessary or the lifetime will be too brief to be useful.

Subsequent to the original design shown in Fig. 2.8, the use of alloys and powder metallurgy has increased greatly the choice of metals used in multielement hollow cathodes. Although their lifetime is short compared to that of a single-element hollow cathode, at times they are very valuable for particular analytical requirements. A recent monograph on hollow cathodes describes their design and function in detail (6).

3. The Monochromator System

The function of the monochromator system is to isolate radiation of the desired wavelength from the rest of the radiation emitted by the light source. The desired radiation, which is usually the emitted resonance line of the element being analyzed, is permitted to travel down the light path to the detector, while all other radiation is prevented by this system from reaching the detector.

The two major components of the monochromator involved in this isolation are the slits and the dispersion element (Fig. 2.9). It can be seen that there are two slits, an entrance slit and an exit slit. The entrance slit permits radiation from the

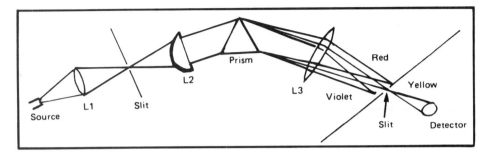

Figure 2.9 System of entrance and exit slits.

hollow cathode to reach the dispersion element, but prevents all other radiation such as stray radiation from lights, windows, etc., from continuing down the light path. The exit slit, on the other hand, is placed *after* the dispersion element, and its function is to permit only radiation of the correct and desired wavelength to pass to the detector. All other radiation is prevented from passing.

In practice, the slits are usually made of two parallel knife edges between which radiation can pass. Often the distance between the edges of the slits is variable, but on some systems this slit width is fixed. In the exit slit the distance between the slits controls the *wavelength range* that is permitted to pass through. The closer together the slits are, the narroweer the permitted wavelength range will be. This wavelength range is called the *spectral slit width*. The physical distance between the jaws of the slide is called the *mechanical slit width*. In atomic absorption spectroscopy the resonance line emitted by the source is very narrow and there are frequently no other lines in the vicinity of this emitted line. Consequently, in many cases low-resolution equipment is all that is necessary. There are cases, however, where high resolution is definitely an advantage, particularly with the transition elements where the emission spectra are very rich. Commercial equipment is designed to operate sufficiently well to meet all resolution requirements.

The function of the dispersion element is to disperse the radiation that falls upon it according to wavelength. It operates in the same manner as a prism that separates the sun's light into a rainbow, as shown in Fig. 2.10.

It will be remembered that the hollow cathode emits a number of very narrow lines and that the background between these lines is virtually zero. It is necessary to separate the resonance line from other spectral lines in its immediate vicinity. These lines originate from the metals in the hollow cathode and the filler gas. The spectra have long been documented, and the wavelengths of these lines are well known.

The analytical *sensitivity* of atomic absorption is defined as that concentration of metal which will diminish the signal by 1%. If the intensity of the signal is I_0

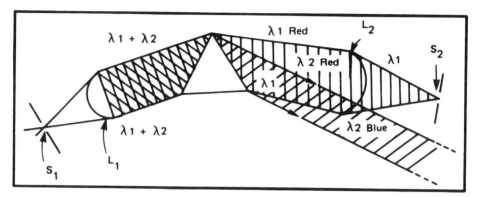

Figure 2.10 Dispersion of radiation into different wavelengths. Focusing lenses are used to control the light. Here the radiation source emits red (long-wavelength) and blue (short-wavelength) light. The light is focused on the entrance slit S_1 and proceeds to collimating lens L_1, which converts it to a parallel beam falling on one face of the prism. The face is completely filled with light. Here the light is dispersed and falls on a second lens L_2, which focuses it onto the exit slit S_2. The latter is placed so that only light of the derived wavelength proceeds to the detector. Other light is blocked out.

before absorption and I_1 after absorption, the <u>sensitivity limit</u> can be defined as

$$\frac{I_0 - I_1}{I_0} = \frac{1}{100} \tag{2.1}$$

If, however, a second line of intensity I_0 falls on the detector and is *not absorbed*, the same concentration that originally gave 1% absorption will now give absorption equal to

$$\frac{(I_0 - I_1) + I_0}{I_0 + I_0} \tag{2.2}$$

The loss of sensitivity depends on the relative intensity of I_0 and I_0. In order to achieve maximum sensitivity it is better to eliminate all spurious sources of radiation. This is achieved by using sufficient resolution to prevent the signal I_0 from falling on the detector.

The principal types of dispersion elements used in atomic absorption spectroscopy are the prism, the grating, and the filter.

a. Required Resolution

The required resolution to separate two lines is defined as

$$R = \frac{Av\lambda}{\Delta\lambda} \tag{2.3}$$

where $Av\lambda$ is the average wavelengths of two lines and $\Delta\lambda$ is the difference in wavelength between the two lines to be resolved. To separate two lines of wavelengths 500.0 and 500.2 nm, the required resolution would be 500.1/0.2 = 2500.5.

b. Prisms

It is not generally possible to use glass prisms in atomic absorption spectroscopy because glass is not transparent at wavelengths less than 350 nm. In practice, therefore, quartz prisms are preferred.

The resolution of a prism is given by the equation:

$$R = \frac{\delta n}{\delta \lambda} t \qquad (2.4)$$

where

δn = rate of change of refractive index
$\delta \lambda$ = rate of change of wavelength
t = thickness of the base of the prism

The rate of change of refractive index with wavelength depends upon the wavelength itself and varies throughout the spectral region. However, most prisms used in atomic absorption are adequate for the work, since only low resolution is usually required.

c. Gratings

A grating consists of a metal surface onto which parallel grooves have been cut, all equidistant from each other. The grooved surface reflects and resolves the light falling upon it. Reinforcement of the light occurs when

$$n\lambda = d(\sin i \pm \sin \theta) \qquad (2.5)$$

where

n = a whole number (the order)
λ = wavelength of absorbed light
d = distance between grooves
i = angle of incidence of light on the grating surface
θ = angle of dispersion of radiation wavelength

The first step in the manufacture of reflection gratings involves the preparation of a master grating from which numerous replica gratings can be formed. The master grating consists of a large number of parallel and closely spaced grooves ruled on a hard, polished surface with a suitably shaped diamond knife. A magnified cross-sectional view of a few typical grooves is shown in Fig. 2.11. For the ultraviolet and visible region, a grating will contain from 300 to 2000 grooves/mm,

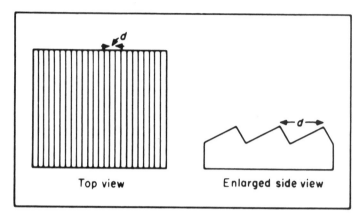

Figure 2.11 Highly magnified view of a grating monochromator.

with 1200 to 1400 being most common. The construction of a good master grating is tedious, time-consuming, and expensive, because the grooves must be identical in size, exactly parallel, and equally spaced over the length of the grating (3–10 cm). The temperature must be rigidly controlled throughout the cutting period to avoid uneven contraction on cooling. However, the process has been automated to the point that it is not an engineering problem but can be a mechanical problem. It is still expensive.

Replica gratings are formed from a master grating by evaporating a film of aluminum onto the latter after it has been coated with a parting agent, which permits ready separation of the aluminum from the master. A glass plate is cemented to the aluminum; the plate and film can then be lifted from the master mold giving the finished replica grating. Replica gratings are much cheaper than originals but the surface is weak and easily damaged. It must be protected at all times and cleaned only with special precautions to avoid damage.

Figure 2.12 is a schematic representation of a grating, which is grooved or "blazed" such that each groove has a relatively broad face, from which reflection occurs, and a narrow unused face. This geometry provides highly efficient diffraction of radiation. Each of the broad faces can be considered to be a point source of radiation; thus interference among the reflected beams 1, 2, and 3 can occur. In order for the interference to be constructive, it is necessary that the path lengths differ by an integral multiple n of the wavelength of the incident beam.

In Figure 2.13, parallel beams of monochromatic radiation X_1 and X_2 are shown striking the grating at an incident angle i to the grating normal. Maximum constructive interference is shown as occurring at the reflected angle θ. It is evident that beam X_2 travels a greater distance than beam X_1 and that this difference is

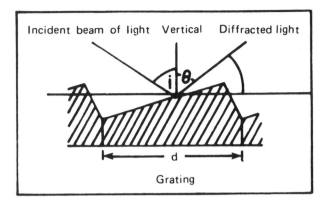

Figure 2.12 Path of light diffracted by a grating.

equal to $(BG - FD)$. Conversely, different wavelengths at different orders can come off at the same angle as shown in Fig. 2.14. This is order overlap and frequently must be eliminated for the best results. An *order sorter* can be used to do this. A low-dispersion prism or a set of filters set before the grating allows only a limited wavelength range to fall on the detector, and overlapping orders are removed. For example, in Fig. 2.15, if the wavelength 600 nm were to be examined, the wavelengths 300 and 200 nm would be eliminated with an order sorter.

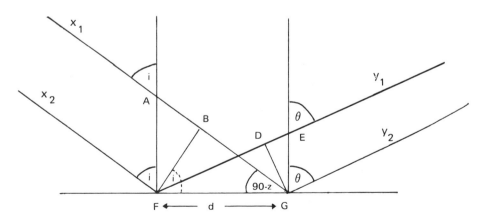

Figure 2.13 Construction of radiation beams for deducing reinforcement and resolution of a grating.

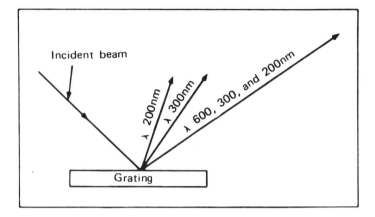

Figure 2.14   Angle of dispersion of light of different wavelengths.

When *d*, the distance between spacings, is decreased, the wavelengths of the operating range of the grating is increased and the working wavelength range is decreased.

In practice, original gratings are expensive and seldom used. A replica grating, made in much the same way that musical records are made, is quite satisfactory and far less expensive. The grating is the most popular type of dispersion element used in atomic absorption spectroscopy for several reasons, the most important of

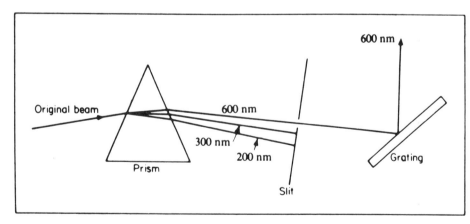

Figure 2.15   Prism used as an order sorter. Note overlapping orders at 300 nm and 200 nm are eliminated.

which is that the grating is durable, its dispersion is constant over its useful spectral range, and suitable gratings are now reasonable in cost.

The resolution of a grating is given by

$$R = Nn \qquad (2.6)$$

Conditions for reinforcement and the resolution of a grating are deduced from Fig. 2.13 as follows:

The beam impinging on the surface is coherent (in phase). When reinforcement occurs, the beam leaving the surface is still coherent (in phase).

Beam $X_1 B = X_2 F$ (Beams are in phase)
$DY_1 = GY_2$ (Beams are in phase)

The two beams $X_1 Y_2$ and $X_2 Y_1$ travel different paths.
Difference in beam length between their paths = $BG - FD$.
Reinforcement when $BG - FD = n\lambda$.

Consider AFG:
  $\angle AFG = 90$
  $\angle FAG = i$
  $\angle AGF = 90 - i$

Consider ΔFBG:
  $\angle AGF = 90 - i$
  $\angle FBG = 90$
  $\angle BFG = i$
   $BG = d \sin i$

Consider ΔEFG:
  $\angle FGE = 90$
  $\angle FEG = \theta$
  $\angle EFG = 90 - \phi$

Consider ΔFDG:
  $\angle EFG = 90 - \theta$
  $\angle FDG = 90$
  $\angle DGF = \upsilon$
   $FD = d \sin \upsilon$

Difference in path length = $BG - FD$
                      = $d \sin i - d \sin \theta$
                      = $d (\sin i - \sin \theta)$

Therefore reinforcement occurs when $n\lambda = d (\sin i - \sin \theta)$ from A.

Where n is the number of waves, it must be a whole number for reinforcement to occur (n is known as the order). If there is one extra wave in path X_2FY_1, $n = 1$. This is first order. If there are two extra waves in the extra distance traveled, $n = 2$ and is second order, etc. Different orders of the same wavelength come off at different angles as shown in Fig. 2.14, where N is the total number of lines ruled on the grating and n is the order. It can be seen that resolution increases with the number of lines ruled on the grating. Suitable commercial gratings are readily available for this work.

d. Holographic Gratings

The holographic grating is not a ruled system, therefore it does not have a blaze angle. It is prepared by utilizing interference fringes generated by two out-of-phase laser beams on photographic material. The fact that the system is essentially flat greatly reduces stray light. This is particularly useful for double monochromator systems where stray light can become a problem. It is fairly common practice to use a ruled grating with a single monochromator but to use holograph gratings with double monochromators.

e. Filter Monochromators

Light filters that pass only a limited spectral range have been used in atomic absorption spectroscopy. One difficulty with this type of filter is that the spectral range it permits to pass is quite wide. The most successful filters are interference filters. Part of the light beam passes directly through and part is reflected by the surfaces of the filters. If the thickness of the filter is d, then when $n\lambda = 2d$, reinforcement takes place and the light is permitted to pass. Light of other wavelengths is lost by interference effects. The system is shown in Fig. 2.16. These filters are available only for the longer wavelength ranges used in atomic absorption spectroscopy.

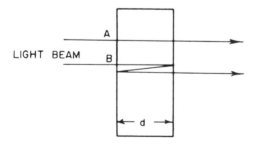

Figure 2.16 Interference filter: Reinforcement occurs when the extra distance traveled ($2d$) by beam B is a whole number of wavelengths ($n\lambda$).

4. Detectors

By far the most common detector used in atomic absorption spectroscopy is the photomultiplier (Fig. 2.17). It can be seen that the photomultiplier is a series of electrodes (or dynodes), each with an emissive surface and a positive potential relative to the previous electrode. When a photon hits the first emissive surface, an electron is ejected and attracted to the next dynode. It is accelerated in the process, and when it hits the next dynode it ejects several electrons. These in turn are attracted to the following dynode, and they in turn each eject several more electrons. This process is continued through each dynode, and a shower of electrons arrives at the final gathering post. In this manner the single photon is responsible for generating a shower of electrons and therefore a significant electrical signal. The sensitivity of this system, which is in fact a simple amplification system, is dependent on the voltage between the dynodes. The greater the voltage difference, the greater the amplification. However, if the voltage is increased too much, the signal becomes erratic and the output becomes "noisy." Typical noise and accceptable signals are shown in Fig. 2.18. In practice, the photomultiplier is operated at the maximum voltage where it is not noisy. This can be determined manually by turning up the voltage until the signal becomes erratic (S/N ratio is too high), and then turning down the voltage until it is just quiet again (S/N ratio is acceptable).

The type of photomultiplier used depends on the wavelength of the radiation

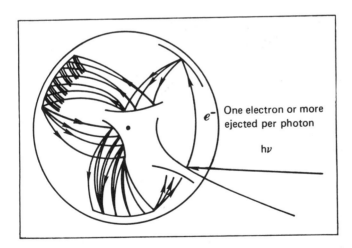

One electron or more ejected per photon

e^-

$h\nu$

Figure 2.17 Photomultiplier. Impinging photons liberate electrons from a light-sensitive metal. The liberated electrons are accelerated to a second electrode, each liberating several electrons on impact.

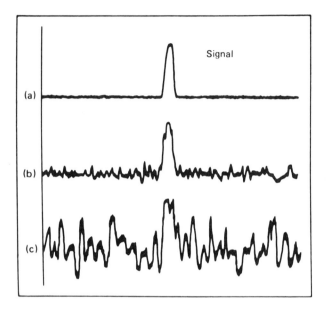

Figure 2.18 Signal versus wavelength with different noise levels: (a) no detectable noise, (b) moderate noise, and (c) high noise.

being monitored. Different photomultipliers use different photoemissive surfaces to permit the energy of the photon to liberate electrons. A change in the surface changes the effective wavelength range over which the photomultiplier responds. The relationship between signal and wavelength for a radiation source of equal intensity is illustrated in a response curve. Typical response curves for the most common photomultipliers are shown in Fig. 2.19.

The output signal varies considerably with wavelength and hence with the photomultiplier used. It is essential that the correct photomultiplier be used, because if the wavelength range of the resonance line is outside the range of the photomultiplier, then insensitive, erratic, and unreliable analytical data will be obtained.

As indicated before, it is also necessary that the photomultiplier be operated at the correct voltage to provide a quiet signal.

5. Atomizer

The function of the atomizer is to reduce the metal analyte to the neutral atomic state. This is a very difficult process to perform reproducibly since it includes a

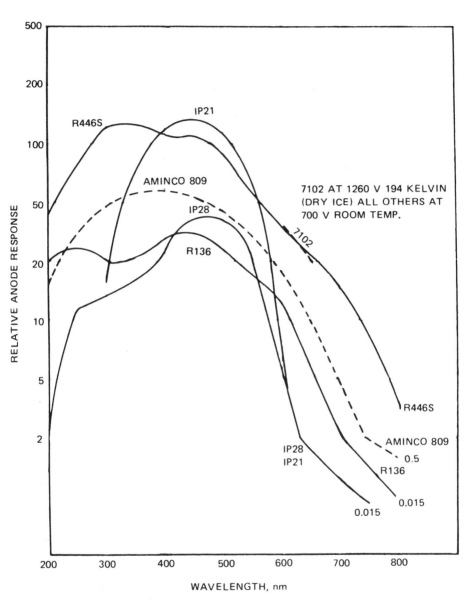

Figure 2.19 Response curves for various types of photomultipliers. Note the variable range between models and the sharp dropoff response outside the usable range.

number of dynamic equilibria each of which is easily displaced. It is at the very heart of the difficulties of atomic absorption spectroscopy. All the other major components of a typical commercial atomic absorption instrument are readily available at a quality level capable of providing maximum sensitivity and reproducibility. In general, atomizers are very inefficient. But reasonably high sensitivities for qualitative detection are readily available using commercial atomizers. However, the problem of providing reproducible quantitative analysis is a more difficult task and involves a true understanding and practice of analytical chemistry. The most common atomizers in routine use are flame atomizers.

a. Flame Atomizers

The process of atomization in a flame will be discussed in Section 5 p. 96. All atomizers have the same general basic functions. A flame is formed using air, oxygen, or nitrous oxide as the oxidant and hydrogen, acetylene, coal gas, or butane as the fuel. The exotic flames such as hydrogen-fluorine or oxy-cyanogen have been used but not on a routine basis. A liquid sample is nebulized to form fine liquid droplets, which are introduced into the base of the flame where the solvent is evaporated. The dissolved sample residue is decomposed and free atoms are generated. The free atoms find themselves in the hostile environment of the flame and are rapidly oxidized either to the metal oxide or to some other more stable molecular form, such as As_4. Their lifetime as a free atom is transient. Flames are very inefficient atomizers. It can be calculated, for instance, that approximately one atom in a million of the sample element is actually reduced to the free atomic state. All other atoms are not reduced to free atoms quickly enough, and they pass through the light path unreduced and do not contribute to the atomic absorption signal. In recent years this problem has been circumvented by using carbon atomizers with greatly increased sensitivity. The design and use of the carbon atomizer is discussed in Section E.1, p. 147.

The sample is introduced as a liquid into the base of the flame of a flame atomizer. Here it is nebulized into small drops, which are evaporated and then atomized as they proceed through the flame itself. The free atoms thus formed find themselves in the hostile environment of a flame and usually are oxidized quickly to some other chemical form. It has been shown on numerous occasions that the efficiency of producing atoms by flames is very low. It is not uncommon to reduce only one atom to the atomic state for every 1,000,000 chemically bound atoms introduced into the flame.

Two main types of burners have been used extensively in atomic absorption spectroscopy: the total-consumption burner and the Lundegardh burner. Of these the Lundegardh burner is much more popular. Each burner is deceptively easy to operate and generates analytical data. However, it is only under very carefully controlled conditions that the data are reliable. Total-consumption burners and Lundegardh burners are illustrated in Figs. 2.20 and 2.21.

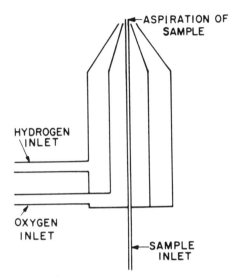

Figure 2.20 Schematic diagram of a total-consumption burner.

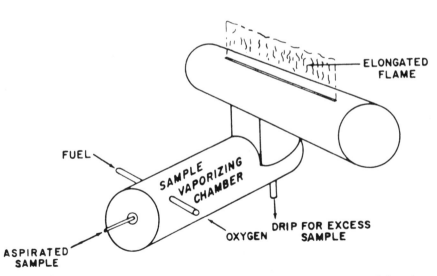

Figure 2.21 A nebulizer burner based on the Lundegardh burner (after the Perkin-Elmer 303).

b. Total-Consumption Burner

The sample is introduced into the base of the flame by aspiration created by the venturi action of the oxidant (e.g., O_2) of the combustion process. The entire sample is introduced to the flame even if it includes suspended particles or large solvent droplets, hence the term "total-consumption" burner. The advantages of the total-consumption burner are as follows: (1) It is comparatively simple to manufacture and, therefore, relatively inexpensive. (2) No fractionation of the sample takes place during aspiration. This eliminates sources of error by loss of nonvolatile components. Its chief disadvantages include the following: (1) The analytical signal changes as the rate of aspiration of the sample into the flame changes. The aspiration rate is a function of the viscosity of the sample. The viscosity changes with temperature changes or solvent. (2) Very viscous samples cannot be aspirated at all and so cannot be analyzed. They may be diluted with a less viscous solvent. (3) After introduction into the flame, the sample is broken into droplets. The droplet size depends on the chemical composition of the sample; that is, any factor that affects surface tension or viscosity will affect the sample droplet size. This in turn affects the atomic absorption signal obtained and therefore the analytical data achieved. Changes in surface tension can cause direct error in analytical data. (4) The burner tip can become encrusted with salts left after evaporation of the solvent. This can cause a change in aspiration rate of the sample and hence a direct error in the analytical results. Special precautions must be taken by the operator to be sure that the burner does not become encrusted during operation. (5) The burners are very noisy, both physically and electronically. The physical noise is very disturbing to the operator and is even more disturbing to other people in the lab. (6) The electronic noise contributes to poor reproducibility in the analytical data. (7) The major part of the sample passes through the flame without evaporating completely (or atomizing) and does not contribute to the signal.

A mechanically fed total-consumption burner with a feed rate independent of viscosity and surface tension was developed (7) that eliminated many of these problems. It will be discussed later.

The total-consumption burner was used extensively in flame photometry but is used less extensively now. Reproducible results can be achieved in atomic absorption spectroscopy, but it is seldom used in atomic absorption anymore.

c. The Lundegardh Burner

In this burner the sample is aspirated into a mixing chamber. Large droplets deposit on the walls and drain off. Small drops and evaporated solvent are swept into the base of the flame where combustion takes place. The practical advantages of this burner are as follows: (1) The elongated flame introduces more atoms into the light path, increasing analytical sensitivity. (2) Burner encrustation is reduced because large sample droplets are eliminated from the system in the mixing chamber. (3) The burner is quiet to operate and therefore reduces irritation to the operator. (4) The analytical signal is significantly less noisy than the total-consumption burner because sample droplets are effectively removed from the light path and light scattering is greatly reduced.

However, the system has some limitations: (1) If the sample contains more than one solvent, any volatile material present is preferentially evaporated in the mixing chamber and less volatile components drain off and do not reach the burner. It is possible under these circumstancces to lose a significant portion of the sample element down the drain rather than sweeping it into the burner, resulting in loss of sensitivity and unacceptable analytical inaccuracy. (2) Sensitivity has been improved by using an impact bead or other spoiler devices. The relatively large volume of the mixing chamber provides a hazard when volatile combustible samples are being analyzed. Special precautions must be taken if oxy-acetylene flames are to be used because of the danger of flashback.

The nitrous oxide-acetylene flame has been used extensively for some applications of atomic absorption spectroscopy. It has been found in practice that the Lundegardh-type burner can be used for this flame quite successfully. Care must be taken to safeguard against flashback of the fuel-oxidant mixture. The manufacturers have done quite a good job in providing suitable safety precautions when the burner is used under these circumstances.

The Boling burner is a modification of the Lundegardh burner. This burner includes three slots instead of one as the orifice for supporting the flame. With this design the entire light beam in the optical system can be enclosed inside the flame, and an increase in sensitivity and accuracy results.

Beckman Corporation has introduced into the market a Lundegardh-type burner in which the nebulizing chamber was heated to 700–800°C. The increased temperature increased nebulization significantly, and an overall increase in analytical sensitivity resulted. However, a memory effect caused problems with quantitative analysis.

A schematic diagram of a recent Perkin-Elmer burner is shown in Fig. 2.22. This type of burner has been successfully interfaced with continuous flow systems. This enables continuous monitoring to be achieved or the injection of a single sample into a continuous flow of the solvent or the effluent from a chromatogram (8). However, the large mixing chamber causes loss of chromatographic resolution.

d. Other Injection Systems

Flow injection has been used for introducing the sample into the atomizer. In this system the solvent is fed into the atomizer continuously and the sample is injected into the flowing solvent. The process is a modification of the injection system used in high-performance liquid chromatography (HPLC). It is particularly useful when only limited sample is available. It also enables exact quantities of sample to be

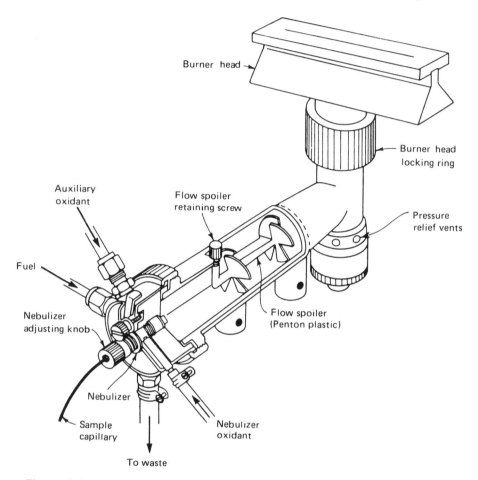

Burner head

Burner head
locking ring

Auxiliary
oxidant

Flow spoiler
retaining screw

Pressure
relief vents

Fuel

Nebulizer
adjusting knob

Flow spoiler
(Penton plastic)

Nebulizer

Sample
capillary

Nebulizer
oxidant

To waste

Figure 2.22 Slot burner and expansion chamber. (Courtesy of Walter Slavin, Perkin-Elmer Corp. Norwalk, Conn.)

injected even when that quantity is small. This permits better quantitation since the volume of sample injected is much more reproducible.

Flow injection systems have also been used for quantitative determination of metals in liquid samples using a computer intelligent flow injection ICP (9). The authors reported improved statistics and reproducibility in the data. A total of 36 elements were determined. It was possible to analyze 76 samples without operator attention.

In summary, flame atomizers have been demonstrated to be very useful and a reliable tool in atomic absorption spectroscopy. They are sufficiently efficient to

Table 2.2 Comparison of Sensitivity[a]

Nebulizer type/metal	Ag	Ca	Cd	Cu	Mg	Zn
Pneumatic	0.15	0.3	0.06	0.4	0.02	0.03
Ultrasonic	—	0.07	0.01	0.07	0.006	0.007
Thermosprayer	0.04	0.04	0.01	0.03	0.003	0.005

[a]1% absorption, 1.5 ml/min.

provide an analytical technique to carry out analyses of the part per million level or lower, and this satisfies most needs. Their atomization efficiency is known to be low and there is usually a rapid loss of atoms from the flame system by oxidation. However, the method has been very reliable for quantitative analysis and is widely used.

Other injection systems include the use of *ultrasonic nebulizers* (10), which gave some improvement in sensitivity but which lost sensitivity because the nebulized droplets coalesced before reaching the burner. The problem was alleviated by heating the droplets (11).

Thermospray nebulizers have also been found to improve sensitivity. It has been used extensively in interfacing liquid chromatography and mass spectroscopy (12) and more recently for interfacing LC and flame atomizer (13).

Sensitivities were improved, as shown in Table 2.2, between one and two orders of magnitude (13). Unfortunately the systems are not commercially available.

6. Modulation of Equipment

Atomic absorption is measured by the absorption of atomic resonance lines of the element being determined. The resonance lines are associated with transitions between the ground state and one of the low excited states of the atom in question. In flame photometry the atoms are excited first to the upper energy levels; then they fall to the ground state and in the process emit radiation. The transitions involved are the same transitions as those involved in absorption, so atoms emit at the resonance line—that is, at exactly the same wavelength at which they absorb. This presents a problem in measuring the absorption by the ground-state atoms. A schematic diagram of the equipment is shown in Fig. 2.23. It can be seen that the intensity of the source starts off at I_0, and after absorption the intensity drops to I_1. The absorption signal, therefore, is $I_0 - I_1$. When the radiant energy *emitted* from the sample is added to this signal, however, the final radiation intensity falling on the detector after absorption is not I_1, but $I_1 + S$, where S is the intensity of radiation emitted at the resonance wavelength.

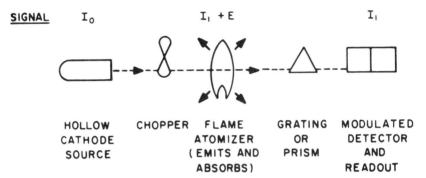

Figure 2.23 Schematic diagram of a single-beam atomic absorption instrument.

This presents two problems. First, there is a significant decrease in sensitivity depending on the intensity of S. The absorption signal is no longer $I_0 - I_1$, but is $I_0 - I_1 + S$.

For many elements, particularly those with resonance lines less than 300 nm, the emission intensity is extremely small, and therefore the error involved in the incorporation of a signal low in intensity S is minimal. But for some elements, such as sodium or potassium, which emit at long wavelengths, the emission signal is large. The net absorption signal $I_0 - I_1 + S$ may be very small or even negative.

The second and more important problem is that S, the intensity of emission observed in flame photometry, is subject to numerous variables. Therefore it is not possible to correct for S based on the concentration of the sample (which is generally unknown anyway) or by using a calibration curve. As a result, an analytical interference results and inaccurate data are obtained.

The problem was ingeniously overcome by Walsh (1), who *modulated* his system. This process involves using an alternating (AC) or interrupted DC signal from the hollow cathode light source and using an A/C detector. All radiation from the flame is DC (or continuous). Since the detector can register only the alternating signal emitted from the hollow cathode, it cannot register the steady signal from the flame. The effect is that the detector reads out only the signal from the radiation source, but does not include the emitted DC radiation from the flame. This eliminates any interference arising from sample radiation.

There are two methods of providing an AC radiation source. One is to use an intermittent electrical impulse to the hollow cathode, which then radiates an intermittent signal. The second method is to use a rotating quadrant in front of the hollow cathode. The latter alternately (1) reflects the signal and (2) allows the signal to pass down the light path. This provides an interrupted radiation signal, one that alternates between zero and I_0. This device is known as a mechanical chopper. In each case the frequency of chopping is maintained at a very controlled rate.

The amplifier used to amplify the signal from the detector is usually a simple AC amplifier, which measures the amplitude of the signal, whatever the frequency of that signal may be. This is illustrated in Figs. 2.24 and 2.25.

These amplifiers work reasonably well under most conditions and are used in most commercial equipment. One problem with these amplifiers is that they will record a signal that is generated by flame flicker. Because they are able to detect any change in signal intensity, they cannot discriminate between an alternating signal coming from the hollow cathode and flicker in the flame that results in small changes in light signal. The net result is analytical inaccuracy, especially if high-intensity emission from the flame is encountered.

If necessary, this problem can be overcome by using a *lock-in amplifier*. The lock-in amplifier is tuned to exactly the same frequency as the hollow cathode itself, and does not pick up flicker or any other AC radiation that occurs at a

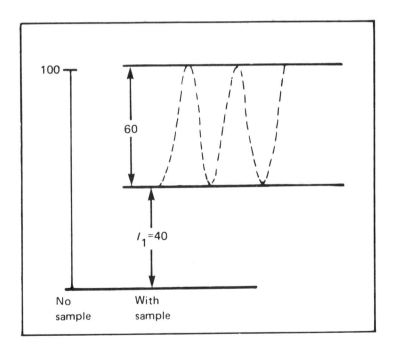

Figure 2.24 Signals obtained with modulation but no emission from the sample. The amplitude of the AC signal is 60, indicating 60% absorption. $I_0 = 100$; $I_1 = 40$.

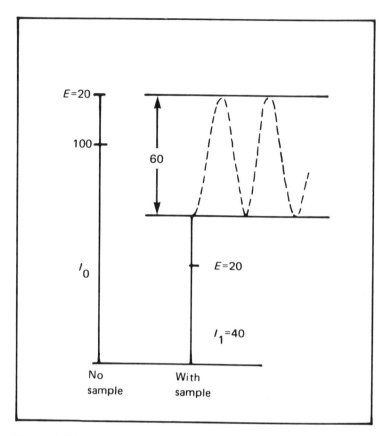

Figure 2.25 Signals obtained with modulation and emission from sample. The amplitude is 60, indicating 60% absorption. The amplitude is independent of emission intensity. $I_0 + E = 120$; $I_1 + E = 60$.

different frequency from that of the hollow cathode. The lock-in amplifier provides better accuracy, if this is a necessity, but naturally increases the cost of the equipment.

A secondary effect of emission from the flame is to fatigue the detector, particularly if the light intensity from the flame is intense. Although the intense radiation from the flame does not contribute to the absorption signal, the detector nevertheless is exposed to this radiation and may become fatigued if the radiation is not eliminated. Failure to do this results in unreliable and erratic data.

7. Correction for Flame Absorption

Molecular species in a flame may absorb at the same wavelength as the atomic absorption line. This introduces a direct source of error. The error can be eliminated by using one of several techniques. The most common are the use of (1) a D_2 lamp, (2) a nonresonance atomic line at a wavelength close to the absorption line, (3) the Zeeman effect, and (4) the Smith-Heifje procedure. These techniques are described later.

C. ANALYTICAL PARAMETERS

1. Choice of Absorption Line

Atomic theory tells us that the electrons in all atoms are in well-defined orbitals, as discussed on page 6. For example, in uranium the electron shells with principal quantum numbers 1 through 6 are all filled and the shell with principal quantum number 7 is partially filled. Numerous orbitals are available in each shell. These are the s, p, d orbitals, etc. In the filled shells, each orbital accommodates an electron.

In the unexcited atom, these electrons reside in the orbitals with the lowest energy level. However, eacch of the upper *empty* orbitals is available to accommodate an electron. During excitation the electron with the lowest energy (the valence electron) moves from its normal low-energy orbital to an orbital with a higher energy. This orbital may be in the same shell or in a higher shell, inasmuch as each orbital is available to accommodate an electron unless excluded by quantum theory–forbidden transitions.

In atomic sodium, electrons fill the shells with quantum numbers 1 and 2, and one electron is in the shell with quantum number 3. When sodium is in the ground state, this will be the orbital with the lowest energy, that is, 3s. If we excite sodium, the 3s electron can move to an orbital with higher energy. The energy level next to the 3s level is the 3p energy level, hence it is possible for an electron to go from a 3s to a 3p orbital. It is also possible for the 3s electron to go into orbitals of even higher energy, such as 4p orbitals, 4d orbitals, 5p, 5d, etc.

When the valence electron of sodium is in the 3s orbital, its lowest energy state, the sodium is said to be in the *ground state*. When the electron is in any orbital with higher energy, the sodium is said to be in an *excited energy state*, or we can say that we have "excited sodium."

An atom can become excited by heating or by absorbing energy of the correct wavelength. The wavelengths of the energies involved are well known and follow standard physical laws, as represented by Eq. (1.3).

When radiation energy is absorbed, the atom becomes excited. If we use a prism to disperse the radiation falling on the atoms, the absorption spectrum appears as

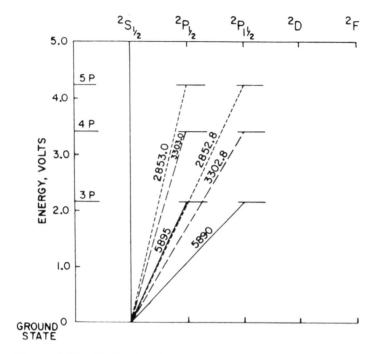

Figure 2.26 Partial Grotrian diagram for sodium.

a series of narrow lines as opposed to wide bands. If the transition is between the ground state and the lowest excited state, then it is said that the absorption line is the *resonance line*.

Transitions between the ground state and upper excited states are possible but are not often used. Often the wavelength of the pertinent absorption level is outside the UV range. The energy levels of sodium are illustrated in the partial Grotrian diagram such as that shown in Fig. 2.26. For the sake of clarity, many of the upper-state transitions are omitted. In sodium the transition between the 3s orbital and a 3p orbital can be achieved by absorbing radiation at 589 or 589.5 nm (the sodium D doublet). Similar absorption of radiation at 33.03 nm will cause sodium to be excited from the 3s ground-state to the 5p excited-state orbital. Transitions between the 3s orbital and orbitals with principal quantum number 6 require more energy. In the case of sodium, such absorption is quite feasible. But with many elements, such as carbon, more energy is required than is available from ultraviolet radiation. These absorption lines exist in the vacuum ultraviolet region which cannot be observed by conventional equipment.

A further disadvantage of the use of these lines is that invariably the oscillator strength is less than that of the resonance line and a significant loss in sensitivity results.

In theory, transition from any excited state to a higher excited state can occur following absorption. In practice, transitions between upper excited states are not observed. In practice, at temperatures encountered in flames or carbon atomizers, the vast majority of atoms exist in the ground state and the electrons are in the lowest energy state as indicated by the Boltzmann distribution. The upper excited states are so sparsely populated that there are too few atoms available for significant percentage absorption of the radiation falling on them. Hence, only transitions that involve the ground state are meaningful.

The most important conclusion is that in atomic absorption spectroscopy the only transitions that are of practical use are between the ground state and an excited state. Absorption is stronger between the ground state and the first excited state (the resonance line), but absorption is possible between the ground state and other higher excited states with reduced sensitivity because invariably the oscillator strength is lower.

For the nonmetallic elements such as the halides, excitation from the ground state to the first excited state requires so much energy that the absorption lines are in the vacuum ultraviolet and cannot be measured with conventional equipment. Therefore, it is not possible to detect directly nonmetallic elements by conventional atomic absorption spectroscopy.

As illustrated with sodium, several absorption lines are sometimes available for use in atomic absorption spectroscopy. This is particularly true of the transition metals. The detection of very small quantities requires the use of the resonance line that absorbs most strongly. However, for higher concentrations absorption lines that absorb less strongly (lower oscillation strength) may be used. The ability to select different absorption lines greatly increases the analytical range of the method.

An illustration is the calibration data for copper using several different absorption lines as shown in Fig. 2.27. The analytical range of the method is extended to 10,000 ppm by simply changing the absorption line.

In summary, the choice of the correct absorption line depends on the analytical sensitivity required. For maximum sensitivity, the resonance line must be used; for lower sensitivity, as required by more concentrated samples, other absorption lines can be used.

It should also be emphasized that the useful absorption lines originate in the ground state. In emission spectrography, emission lines occur as a result of an electron dropping from an excited state to a lower excited state or to the ground state. The most persistent lines are those originating from transitions between low excited states and ground states. These are always the most intense lines, and the last lines to disappear as metal concentration decreases. In practice, the strongest absorption lines are frequently the most intense emission lines, but they self-

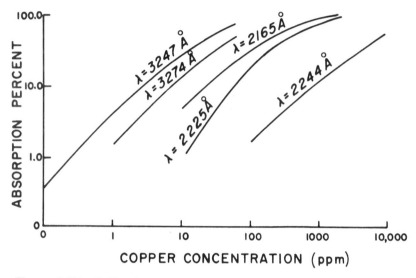

Figure 2.27 Calibration curves for copper using various absorption lines.

absorb and are the lines to avoid for quantitative analysis in emission spectrography except where maximum sensitivity is needed. In contrast they are often used in plasma emission where self-absorption is minimal.

2. Degree of Absorption

The ultimate objective of an analytical determination is to obtain accurate and precise results. In order to do this, it is vital that we understand clearly what is being measured and the effect of variables on this measurement.

The degree of absorption is

$$\int_{0}^{\infty} K v \, dv = \frac{\pi e^2}{mc} \, Nf \tag{2.7}$$

where

$Kvdv$ = total amount of light absorbed at frequency v
π, e, m, c = constants
N = total number of absorbing atoms in the light path
f = oscillator strength of the transition

The total degree of light absorbed, therefore, depends on a number of constants times N, the number of atoms in the light path, and f, the oscillator strength. The oscillator strength depends on the probability of the transition between the ground state and the higher excited state. The greater the probability, the greater the oscillator strength. This is a physical property of the atom and does not vary under normal circumstances. We can see, therefore, that the total degree of absorption depends on the product of a number of constants times the number of atoms in the light path.

The number of atoms formed is equal to the number of atoms of the element in the original sample times the efficiency of atomization. This number is modified by the fact that the free atoms are continuously lost as they react in the atomizer to become oxides or other chemical entities. The number of free atoms N is in dynamic equilibrium between the number created and the number lost:

$$\text{Sample molecules} \longrightarrow \text{free atoms} \rightleftharpoons \text{metal oxides}$$

In order to get reproducible analytical data, it is vital to control both the efficiency of producing atoms and the rate at which the atoms are lost. When this control is effective, dynamic equilibrium is achieved and the number of free atoms in the system remains constant for a given concentration of sample.

Any variable that affects the efficiency of atomization or the rate of loss of free atoms after atomization directly affects N, which in turn controls the absorption measurement and, therefore, directly affects the analytical data. For reproducible results it is necessary to control these variables, which depend to a large degree on the atomizer being used. In this section we shall consider only the flame atomizers, inasmuch as they are by far the most important atomizers used and the most generally available commercially.

3. The Observed Atom Population Profile in the Flame

The normal height of the flame from base to tip of a total-consumption burner is about 3 in. If the atomic absorption is measured at different heights in the flame, it is found to vary considerably. The relationship between height in the flame and atomic absorption signal is called the *flame profile*. Profiles of three different elements are shown in Fig. 2.28 and 2.29, in which it can be seen that the absorption is very low at the base of the flame, increases rapidly to a maximum at the reaction zone, and then decreases slowly as the top of the flame is reached.

The degree of absorption is a direct measure of N, the number of atoms in the light path, and therefore the flame profile is a measure of the atomic population in various parts of the flame. It should be noted that the emission maximum takes place at a low part of the flame, presumably because the lifetime of an excited atom is quite short (10^{-6}–10^{-7} sec) and their population does not accumulate with time. But free atoms have an extended lifetime provided they do not react chemically or become absorbed on the surface. Hence, an accumulation of free atoms occurs as we move into the upper regions of the flame.

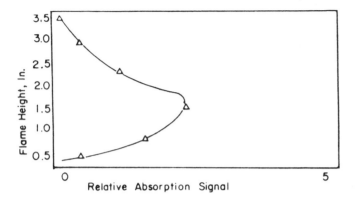

Figure 2.28   Flame profile for nickel, 241.4 nm: relative absorption vs. flame height.

4. The Effect of Different Solvents

The solvent has several direct and indirect effects on the absorption signal. These effects are caused by (1) spectral emission by the solvent or combustion products of the solvent when introduced into the flame, (2) spectral absorption by the solvent or its combustion products in the flame, (3) the effect of the solvent on the efficiency of producing neutral atoms from the sample, (4) the change in sample feed rate with solvents with different viscosities, and (5) the effect of surface tension on drop size when the sample is introduced into the flame.

It has been observed on numerous occasions in both flame photometry and atomic absorption spectroscopy that enhancement of the signal takes place if an organic solvent is used instead of an aqueous solvent. The enhancement of the emission signal and the absorption signal are not equal to each other, but it can be stated unequivocally that in each case a significant enhancement is observed (Table 2.3).

The most significant difference between an organic and an aqueous solvent is the combustibility of the sample (7). If an organic solvent is introduced into a flame, it burns readily, giving off thermal energy—that is, it is an exothermic reaction. On the other hand, if an aqueous sample is introduced into a flame, it requires energy to cause it to evaporate. The latter is an endothermic reaction, which proceeds less efficiently than the organic exothermic reaction (see Table 2.4).

It should be remembered that in atomic absorption the signal depends on the number of free atoms in the ground state. This number is very close to the total number of atoms in the system. However, in flame emission the excitation signal depends on the number of atoms in the particular excited state from which

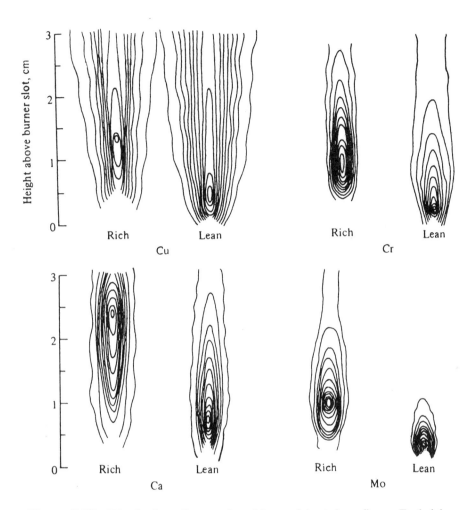

Figure 2.29 Distribution of atoms in a 10-cm air/acetylene flame. Fuel-rich and fuel-lean results are shown. Maximum absorbance is at the center. From Ref. 14.

Table 2.3 Absorption and Emission of Nickel 341.4 nm Resonance Line in Various Solvents (10 ppm Ni)

Solvent	Mechanical-feed burner		Beckman aspiration burner			Viscosity at 20°C (1000 n)
	Absorption 100 - I	Emission	Total absorption	Absorption corrected for feed rate	Emission	
Acetone	14.5	33	15	7.0	144	3.3
n-Heptane	15.5	40	—	—	66	4.2
Ethyl acetate	13.0	37	12	7.5		4.5
Methyl alcohol	15.0	29	9	5.8		5.9
Benzene	16.0	42	2	1.9		6.5
Toluene	15.0	38	16	15.5	30	7.7
Methyl cyclohexane	16.0	39	20	14.0		7.7
Amyl acetate	13.5	36	—	—		8.0
Cyclohexane	15.0	35	8	7.3		9.3
Carbon tetrachloride	8.0	5	5	4.0		9.6
Ethyl alcohol	15.0	31	6	3.7		12.0
Nitrobenzene	13.5	34	2	1.0		19.8
Isopropanol	14.5	33	—	—		22.5
Varsol	14.5	30	5	9.3		—
Water	4.0	4	3	3.0	4	10.0

Source: Ref. 7.

Table 2.4 Solvent Combustion Mechanism

Combustion of an Aqueous Solvent:

	step 1		step 2		step 3
Ions in water	→	Evaporation leaving a hydrated residue	→	Residue (dehydrated atoms formed)	→	Excited atoms (emission) Neutral atoms (absorption)

Combustion of an Organic Solvent:

	step A		step B		step C
Metals in organic solvent	→	Solvent burns	→	Residue organic addend burns	→	Excited atoms (emission) Neutral atoms (absorption)

Steps 1,2,3 are endothermic—low efficiency.
Steps A,B,C are exothermic—high efficiency.

radiation occurs. This number will be very dependent on other factors besides the efficiency of producing atoms. Hence, although we shall observe an increase in both flame emission and absorption, there will not be a direct relationship between the two signals.

There is significant evidence that the enhancement effect is not caused by an increase in flame temperature when an organic solvent is used. Such temperature increase is bound to take place and some enhancement will occur, based on the Boltzmann distribution, but it will not be in the same order of magnitude as that observed in practice.

a. Change in Sample Feed Rate When Solvents Having Different Viscosities Are Used

When a fluid solvent is used, the sample is aspirated into the base of the flame much more rapidly than when a viscous sample is used. This directly affects the sample feed rate and therefore affects the absorption signal. Commercial burners are designed to introduce the sample into the base of the flame. Usually it is assumed that an aqueous sample will be used. The use of organic solvents changes the characteristics considerably. The increased flow rate may cause overloading of the sample, or a decreased flow rate due to a viscous sample may cause the sample injection rate to be significantly reduced.

A mechanical burner has been designed that injected the sample at a fixed flow rate independent of its viscosity and surface tension (7). It was shown that using this burner a constant metal concentration in any organic solvent gave approximately the same absorption signal and that there was virtually no difference between highly viscous samples and highly fluid solvents. The conclusion that could be drawn from these data was that the viscosity per se did not affect the signal significantly, but that it did affect the flow rate into the sample and hence the absorption signal if no correction was made (Table 2.3).

b. Effect of Surface Tension

If a sample has a high surface tension, then large drops of the sample are quite stable. If, on the other hand, the surface tension is low, then only small drops are stable. It can be readily understood that the drop size of the solvent after introduction into the base of the burner will depend very much on the surface tension of the solvent being used. This problem was studied using the mechanical burner mentioned previously. In this instance the sample was broken up by jet force to produce drops of approximately the same size in each case independent of the viscosity of the solvent. The data showed that the absorption signal was independent of the surface tension when the drop size was equalized. This showed that the surface tension per se was not an important attribute to the absorption signal, but that it did affect the drop size and therefore did affect the absorption signal obtained in practice.

Data obtained using the mechanical flow rate burner together with data obtained from a simple Beckman total-consumption burner are shown in Table 2.2. The data indicate that the absorption signal was virtually independent of the viscosity of the sample when the force-feed burner was used. The use of noncombustible solvents such as water and carbon tetrachloride produced a significant change, indicating that combustibility was a factor in producing free atoms. With the total-consumption burner there was a tendency for the absorption signal to decrease as viscosity increased. This tendency was less apparent when a correction for simple feed rate was made. For the samples tested the results were perhaps more noteworthy for the lack of correlation between absorption and viscosity.

It can be concluded that the solvent has a very dramatic effect on both emission and absorption signals. The primary effect is on combustibility of the solvent. However, secondary effects caused by viscosity and surface tension were also noted, particularly when aspiration burners were used.

The most important step to be taken to correct for this problem is to be sure that the solvent of the standards used in making the calibration curve and the solvent of the sample are the same.

5. Factors Affecting Atomization Efficiency

Generally, the sample enters the flame base as a liquid droplet, and the metal of interest leaves it in the form of an oxide. Numerous steps are involved in between (Table 2.5). It will be appreciated, of course, that many of the larger droplets never evaporate completely and therefore pass through the flame virtually unaffected,

Table 2.5 Factors Affecting Flame Profiles

Physical form of sample in flame	Reaction	Factors controlling reaction	Part of flame
Oxide ↑	No reaction or reduction	Stability of metal oxide, flame composition	Outer mantle
Atoms ↑	Accumulation or oxidation	Flame composition, stability of atoms	Reaction zone
Solid particles ↑	Disintegration	Stability of compound, anions, flame temperature, ultraviolet light emitted from the flame	Inner cone
Droplets	Evaporation	Droplet size, solvent, flame temperature feed rate, combustibility	Base

taking with them atoms of the element being determined. Also, quite often the residue left behind after evaporation passes through the flame without further decomposition to free atoms. In each case the atoms contained in the unevaporated droplets or the undecomposed residues pass through the flame and never reach the atomic state. Consequently, they never contribute to the absorption signal.

There are two major segments of the flame absorption profile: (1) At the base of the flame, where the absorption signal increases, atoms are formed and accumulated. (2) In the upper regions of the flame, where the signal decreases, the atoms are oxidized and no longer absorb strongly at the atomic absorption lines. Various factors affect the formation and loss of atoms from the system. The more important ones are as follows.

a. Droplet Size
The sample is aspirated into the base of the flame by the burner. Here it is broken down into small droplets, which pass through the flame. If the droplets are relatively large, their surface area per unit weight is small and evaporation is slow. It is quite possible under these conditions that they will pass through the flame without completely evaporating, and that any metal contained in them are never reduced to the free atomic state and never contribute to the absorption signal. On the other hand, if the drop size is small, then the surface area is relatively large per unit weight and evaporation is rapid. The droplet evaporates, leaving a solid residue which is then decomposed by the thermal energy of the flame. Any metal in the residue is available for reduction to the free atomic state and can contribute to the absorption signal. Most total-consumption burners produce large-, inter-

mediate-, and small-sized droplets, and thereby operate at a usable but not optimum efficiency. In contrast, the Lundegardh burner produces very small drops or even residual particles which are swept by the gas flow into the base of the flame. Its efficiency in atomization is therefore enhanced by this step. On the other hand, up to 90% of the nebulized sample drains off and never reaches the base of the flame.

A second important consideration related directly to drop size is the size of the solid residue remaining after evaporation of the drop. If the size is large, then the residue will be relatively bulky, and decomposition by the thermal energy of the flame will be slow and inefficient. This again will cause a loss of sensitivity, because any residue that is not decomposed cannot contribute to the atomic absorption signal. Large residues can be caused by a high salt content in the sample or by the evaporation of large drops by a hot flame.

Drop size is very dependent on the design of the nebulizer section of the burner. The three most important factors controlling nebulize efficiency are as follows: (1) Maximum sample breakup (minimum drop size) occurs when the sample is introduced into a point of maximum gas velocity. The greater the gas velocity, the lower the particle size; (2) an obstruction in the gas carrying the sample droplets causes a decrease in drop; (3) an abrupt change in pressure, velocity, or direction of the gases—such as at a shock front—will break up the drops to smaller size.

These factors are usually taken into account by commercial manufacturers of burners. The burner design has already been optimized by the manufacturer, and the operator is not called upon to change the design of the burner to be used except under research conditions.

b. Chemical Stability of the Residue

Another factor affecting the efficiency of decomposition of the residue is the chemical stability of the metal salts involved. For example, suppose we have two aqueous solutions of aluminum, each with a concentration of 10 parts per million. One solution is aluminum chloride, and the second solution is aluminum hydroxide. When the solutions are introduced into the base of the flame, the solvent evaporates off and we are left with a residue of equal size including aluminum chloride in one case and aluminum hydroxide in the second case. The aluminum chloride will be more easily broken down than aluminum hydroxide to produce free aluminum atoms. Hence, we would get a stronger absorption signal from a solution of aluminum chloride. This phenomenon is at the heart of chemical interference and is a major source of error in atomic absorption spectroscopy. It will be discussed in Sec. 3.11.

c. Sample Feed Rate

Droplet evaporation and residue disintegration require energy, which is provided by the flame in the total-consumption burner. In the Lundegardh burner and the preheat burner, the evaporation takes place in the mixing chamber and relieves the flame of some of this function. Under controlled conditions the flame produces energy at a steady rate. If the sample feed ratee is too high, too much of the energy of the flame is used up in nebulizing and disintegrating the sample. The final atomization step is therefore inefficient. Under these conditions the flame is swamped by the sample. On the other hand, if the sample feed rate is too low, the production of neutral atoms is reduced and again the absorption signal is diminished. Between these extremes there is an optimum feed rate, which varies with the burner design and even between different burners of the same design. For reproducible results the sample feed rate must be optimized and kept constant.

As can be readily imagined, change in the solvent causes a significant change in the droplet size because of viscosity changes in the sample. Some burners are adjustable and corrections can be made for such changes, but this is not usually the case. Whenever possible, the burner should be optimized for use. In all cases the conditions used for making up calibration curves should be the same as those used for analyzing the sample. If a different solvent is used in the preparation of the calibration curve than is used in the sample, the data will not be relevant and inaccurate results will be obtained.

d. The Effect of Flame Temperature

The transformation from a solution to free neutral atoms is achieved by absorbing energy from the flame. The energy of the flame is directly related to its temperature.

The hotter the flame, the more efficient it will be in reducing the sample to free atoms. The efficiency of production is related to drop size and to the reduction of free atoms in the residues formed. With low-temperature flames, small changes in the chemical energy required to bring about reduction to atoms can cause significant changes in efficiency of atomization. But with high-temperature flames the efficiency is much greater, and similar small changes in chemical energy have a reduced effect on atomization efficiency. It should be pointed out, however, that if the amount of energy required to break down two different chemical forms is great, then there is a significant change in atomization efficiency even when hot flames are used. These energy changes are brought about by the chemical composition of the sample and are the basis of chemical interference.

The temperatures of typical flames are shown in Table 2.6. Flames commonly used in atomic absorption spectroscopy include air-acetylene, nitrous oxide-acetylene, oxyacetylene, oxyhydrogen, and propane-oxygen.

An exception is when easily ionizable metals such as sodium, potassium, or lithium are to be analyzed. Too high a flame temperature increases ionization and therefore depletes the free atom population. These elements are usually analyzed in low-temperature flames.

e. Oxide Formation

After metals in the sample have been reduced to neutral atoms, they will stay in this state for varying periods of time. This time period ends if and when the atom

Table 2.6 Flame Temperatures of Typical
Flames

Fuel	Oxidant	Flame temperature (°C)
H_2	O_2	2800
H_2	Air	2100
H_2	Ar	1600
Acetylene	O_2	3000
Acetylene	Air	2200
Acetylene	N_2O	3000
Propane	O_2	2800
Propane	Air	1900

reacts with the oxidant or other components of the flame. Oxide formation is more likely in an oxidizing than in a reducing flame.

The ease with which a metal atom is oxidized depends on the chemistry of the particular metal. If the oxide is easily formed and is stable, it will oxidize very easily. The step is controlled by equilibrium:

$$\text{metal} + \text{oxidant} \xrightarrow{K_{eq}} \text{metal oxide}$$

It can be calculated that K_{eq} at 3000°C for silver is about $10^{-2.7}$ and for magnesium about $10^{1.9}$. Their respective flame profiles are shown in Fig. 2.30 and Table 2.7. It can be seen that there is a marked difference between the profiles of magnesium and silver. The magnesium absorption goes through a sharp maximum at the reaction zone of the flame. Optimum sensitivity is obtained only when the light path passes through this maximum. For reproducible quantitative results it is

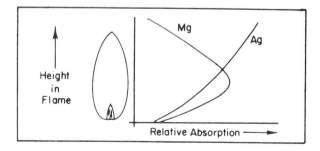

Figure 2.30 Effect of oxide stability on flame profile. MgO = stable oxide; Ag_2O = weak oxide.

Table 2.7 Flame Absorption Profile for Silver and Magnesium

Metal	Flame height for maximum absorption	Predicted stability of oxide
Mg	Medium	High
Ag	High	Medium

important that the light path always passes through the same part of the flame, both when calibration data are obtained and when the samples are being analyzed. It must be remembered that the total height of the flame may vary somewhat from day to day and with changes of fuel/oxidant flow rate, and the points of absorption maximum will vary accordingly. It is therefore important that the absorption maximum be located frequently by actual measurement.

It can be seen that the silver absorption curve does not go through a maximum. This is because silver oxide is easily decomposed and does not form in the flame. Consequently, an accumulation of neutral silver atoms takes place in the higher parts of the flame. This type of flame profile is common with the noble metals.

f. Effect of Variation in the Ratio of Fuel to Oxygen in the Flame

The maximum temperature of a simple flame such as an oxhydrogen flame is achieved when the ratio of oxygen to hydrogen is the stoichiometric ratio necessary for complete combustion of each component. When there is an excess of oxygen, the flame temperature drops and there is an excess of hot hydrogen in the flame. Excess oxygen produces an oxidizing flame, and excess hydrogen produces a reducing flame.

It must also be remembered that in such flames there is an abundance of hot water molecules, moieties such as OH, O_2H, and other reactive chemical entities. These are all capable of oxidizing neutral atoms even in reducing flames.

We can see that the oxidant: fuel ratio directly affects the flame temperature and the lifetime of neutral atoms in the flame. It therefore directly affects the absorption signal that will be obtained from a given sample. For reproducible results the ratio of oxidant to fuel in the flame must be kept constant. For most metals these ratios have been worked out and are published in the literature. Recommended flow rates are also suggested by burner manufacturers.

Independently and simultaneously, Amos (15) and Willis (16) investigated nitrous oxide-acetylene flames. These flames are very hot and require the use of a laminar flow burner to assure maximum atomization. The use of nitrous oxide instead of oxygen as the oxidant greatly reduces the probability of a metal atom becoming oxidized in the flame. Results have shown that they are particularly

Table 2.8 Metals with Refractory Oxides
Reduced in $N_2O-C_2H_2$

Metal	Sensitivity (ppm)
Al	1.0
Be	0.1
Sc	1.0
W	1.0
U	100.0
V	1.0
Zr	50.0

useful for atomizing metals that form refractory oxides. In other flames these metals are reduced only with great difficulty. A partial list of such metals and their sensitivities is given in Table 2.8.

The nitrous oxide-acetylene flame has been shown to have applications for other metals as well as those that are difficult to reduce. It is therefore universal in application and is used extensively for atomic absorption spectroscopy.

g. Cations

In general the presence of several cations in the sample has no effect whatever on the absorption signal of the metal of interest. It has been noted, however, that if the concentration of one interfering cation is very high and its resonance line is very close to the resonance line of the element being determined, then some apparent interference can take place because of absorption of the "wings" of the resonance line of the concentrated metal. This case is unusual and in the majority of instances is of academic interest only.

It can be shown that even different isotopes of the same metal do not absorb at the same resonance line. This has been demonstrated for lithium-6 and lithium-7 and for uranium-235 and uranium-238. Isotope analysis can therefore be achieved by using a lithium-6 hollow cathode lamp as the radiation source and determining the amount of lithium-6 in the sample. No absorption by lithium-7 takes place.

Mutual interference has been noted in the case of magnesium and aluminum. When these two metals are present in the same solution and are aspirated into the flame, there is a decrease in the absorption signal compared to that which would be obtained if only one element were present. This might be explained by the formation of an intermetallic compound, which is not decomposed in the flame. A more likely explanation is the formation of metal oxide salts such a magnesium aluminate. These compounds are difficult to decompose, and their formation would result in a decrease in the formation of both magnesium and aluminum atoms in the flame.

If the sample is known to contain varying amounts of these elements, the problem can be overcome somewhat by using the standard method of addition for calibrating the system.

Although these cases of cationic interference have been observed, there are very few other cases reported of this type of interference. In general, therefore, atomic absorption spectroscopy is free from cationic interference.

h. Anions (Chemical Interference)

Normally in an aqueous solution the metal is in the form of a cation and is dissolved with a corresponding anion. When the solution is aspirated into the flame, droplets are formed which evaporate and leave the residue. The residue contains the metal cation and anion combined to form a molecule. In order to produce neutral atoms from this molecule, the metal-anion bond must first be broken. The strength of this bond determines how easily the step is taken. A strong metal-anion bond is difficult to break and, consequently, for a given flame system production of free metal atoms is less efficient. If, however, the metal-anion is a weak bond, then it will more easily be broken and production of the free atoms will be more easily accomplished. In the latter case, the increased number of free atoms will give a greater atomic absorption signal even though the concentration of the metal in the solution is constant.

If the metal is combined with an organic anion, the situation is somewhat modified but is still similar. The organic anion will burn in the flame, and the free metal atom is liberated quite easily. This results in a greater absorption signal and corresponding increase in sensitivity. It can be seen, therefore, that the absorption signal will depend on the molecular form of the element being determined. An illustration of this chemical effect is shown in Table 2.9. The effect is to change the metal compound form in the absorption signal. This effect can be observed by simply adding different anions to a solution, particularly if these anions are in a large excess compared to the metal being analyzed. For example, a solution of

Table 2.9 The Effect on the Absorption Signal of Changing the Metal Compound Form

Compound	Solvent	Chromium absorption signal, $I_0 - I$ ($I_0 = 100$)
Chromic nitrate	Ethanol	20.5
Sodium chromate	Ethanol	18.2
Chromium naphthenate	Ethanol	26.3

Table 2.10 Chemical Interference and Its Elimination

Interfering anion (100 ppm)	Absorption (%)	
	Lead nitrate solution, plus interfering anion	Lead solution plus anion plus EDTA (1%)
None	23.0	23.0
PO^3_4	19.0	22.8
Cl^-	22.3	22.3
CO^{2-}_3	12.8	23.0
I^-	13.7	23.5
SO^{2-}_4	22.0	23.0
F^-	21.3	22.9

Conditions: Sample, lead nitrate solution, 1.0 ppm; wavelength, 217 nm; flame, oxyhydrogen 30 liters/min:50 liters/min; burner, Beckman total-consumption; aqueous solution.

lead nitrate may be made up such that its concentration is one part per million; but if phosphate ion is added to the solution at a concentration of 100 parts per million, when the solution goes to the normal atomization step there is a high probability that the lead will form lead phosphate rather than nitrate in the residue. In order to achieve atomization it is necessary to break the lead phosphate bond rather than the lead nitrate bond. A summary of this effect is shown in Table 2.10.

The effect can be greatly reduced or even eliminated by adding a strong organic complexing agent. For example, by adding EDTA in the example shown in Table 2.10, the lead will form the lead EDTA complex because the latter complex is more stable than the simple ionic solutions. Under these circumstances the lead will exist in the residue state preferably as the EDTA complex. Therefore, change in signal due to the presence of different anions will be eliminated. It can be seen that the signal remains about the same, irrespective of the predominant anion present in the solution. This is a useful way to eliminate anion effect when samples of unknown composition are being analyzed.

Chemical interference is the most important source of error in atomic absorption.

i. Matrix Effect

It has been observed that the presence of high concentration of salt (e.g., several percent) results in a decrease in the absorption signal. Typical samples that have this kind of high salt concentration include seawater, urine, and other body fluids. The problem arises in the atomization step. After the solvent has been evaporated, we are left with a residue. If the sample has a high salt content, then the residue will be composed mainly of the salt from the sample. The element being analyzed will be entrapped in the comparatively large quantity of residue formed. As the residue goes through the flame, a lot of energy is required to break it down very quickly and liberate the free atoms in order to effect absorption. The efficiency of this process is low, and so it is quite possible for the sample element to pass through the flame remaining in the residue and never be reduced to free atomic state. The net result is a significant loss of sensitivity of the analytical method.

One of the most successful methods for eliminating this problem is simple dilution of the sample, which cuts down the concentration of the salt present. This may not always be possible, particularly if the sample element being determined is already low in concentration. A second successful method is to use solvent extraction, which can be used either the extract the metal of interest or to extract the salt, leaving the metal of interest in an analyzable solution.

Another problem with high-salt-content samples is that they frequently tend to clog the burner, causing a decrease in sample flow rate and a loss of signal. A further problem is that the small residue particles introduced into the flame can scatter the light. To the detector this scattered radiation acts exactly the same as absorbed radiation. However, this can be overcome by dilution of the sample or by using absorption background correction.

6. Spectral Emission from the Solvent and Solvent Combustion Products in the Flame

If we examine the emission signal from a simple oxyhydrogen or oxyacetylene flame, we shall observe several strong emission bands. If we then aspirate aqueous or organic solvents into the flame, we shall find that more emission bands appear. The most common emission bands found are those of the OH radicals, which are formed either when water is introduced into a flame or when organic compounds containing hydrogen are burned, thus producing water in the flame. The mechanism of production of these emission bands will not be discussed in this book; they have been thoroughly described by Gaydon (17). The emission does not represent a problem because it can be eliminated as an interferent by modulating the system. As in atomic emission, any molecular emission signal is DC and does not register when the system is modulated. Trouble can arise if the emission signal is very intense, as in the case of an oxyacetylene flame. When intense radiation falls on the detector, the latter becomes fatigued and erratic, thus producing unreliable data.

7. The Absorption of Spectral Energy by the Flame and Solvents

Simple oxyhydrogen flames are absorbed over an extensive spectral range. The typical absorption curve is shown in Fig. 2.31. The absorption is probably due to OH or O_2H moieties present in the flames. It can be seen that it is quite intense

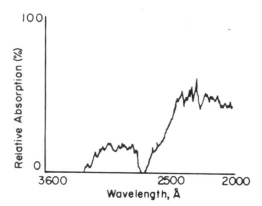

Figure 2.31   Absorption spectrum of oxyhydrogen flame.

over the region 300–320 nm. Similarly, absorption bands are observed when organic solvents are introduced into the flame. The degree of absorption is greater when solvents that contain halides such as CCl_4 are used. This is particularly true at wavelengths less than 300 nm. This is molecular absorption, and although it is not strong it is still not negligible and can cause serious error if corrections are not made. Inasmuch as these are molecular absorption bands, they are not continuous, but are a series of closely adjacent absorption lines. Table 2.11 gives a list of some spectral lines that are absorbed and a second list of lines in the same region that are not absorbed.

The molecular absorption varies significantly with the part of the flame being examined. It is important, therefore, that all absorption measurements be taken at the same part of the flame in order for the correction to be valid. There are several ways to correct for this source of error.

8. Methods Used for Background Correction

a. Using a Blank

Historically the first method used was to measure the absorption of the metal resonance line by the flame and the pure solvent alone. The sample is then analyzed measuring the total absorption of sample element plus the flame system. The difference between the two measurements was the absorption by the sample element being measured.

b. Using a D_2 Lamp

A second method of correction for molecular absorption is to use a deuterium lamp in conjunction with the hollow cathode lamp. Let us suppose that we wish to

Table 2.11 Nonabsorbed Lines for Background Correction

Element	Wavelength of hollow cathode line absorbed (nm)	Wavelength of hollow cathode line not absorbed (nm)
Aluminum	309.2	307.0
Antimony	217.6	217.9
Cadmium	228.8	227.6
Chromium	357.9	352.0
Cobalt	240.7	238.8
Copper	324.7	323.4
Indium	304.0	305.7
Iron	248.3	247.2
Lead	283.3	282.0
Magnesium	285.2	281.7
Molybdenum	313.3	323.4
Nickel	232.0	231.6
Platinum	265.9	270.2
Tin	286.3	283.9
Vanadium	318.4	312.5
Zinc	213.9	212.5

correct for molecular absorption while measuring the atomic absorption of magnesium. Usinq a Mg hollow cathode light source the system would be tuned to the wavelength of the magnesium resonance line 283.3 nm. Set the spectral slit width to 0.2 nm. Here the absorption by the sample would be measured. This signal is the sum of the atomic absorption by magnesium and the molecular absorption by any molecular species that absorb in this spectral region. Let us now replace the hollow cathode with a deuterium lamp, but leave the monochromator at the same wavelength setting. The molecular absorption over the 0.2 nm spectral range will remain as before. But the atomic absorption will remove radiation only from a narrow line (0.001 nm) from the total bandwidth of 0.2 nm. The total light absorbed from the hydrogen lamp will be the molecular absorption over the entire bandwidth. Even if the atomic absorption succeeded in removing all the radiation from a bandwidth of 0.001 nm, the decrease in signal would only be 0.5%. Absorption of the hydrogen lamp by the magnesium is very small and can be ignored. Hence, the molecular absorption signal can be measured by simply measuring the absorption from a hydrogen lamp at the same wavelength as that in which the resonance line occurs. This is illustrated in Fig. 2.32. The difference between the absorption of the hollow cathode line (atomic absorption + molecular absorption) and the deuterium radiation (molecular absorption only) is the net atomic absorption.

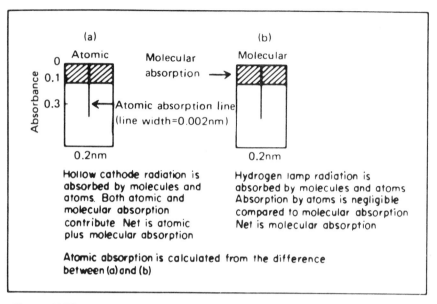

Figure 2.32   Correction for molecular absorption using a hydrogen lamp.

c. Zeeman Background Correction (8.9)

When an element such as mercury emits or absorbs radiation at a resonance line (e.g., 253.3 nm for Hg), the "absorption line" actually consists of several absorption lines, which are very close together and are characteristic of each isotope of mercury. If we direct our attention to one isotope, such as ^{204}Hg, it emits and absorbs at only one characteristic wavelength.

However, if we place the Hg lamp source in a magnetic field, the single line is split into two components by the Zeeman effect (18). The shift is about 10^{-5} nm as illustrated in Fig. 2.33.

Typically the strength of the magnetic fields are between 7 and 15 kgauss. The magnetic field has two important effects on the emission line. In the simplest case it causes splitting into two components: π component at the same wavelength as before and a σ component, which undergoes a positive and negative shift in wavelength. The magnetic field also polarizes the light so the π component goes in one direction, the σ component in the opposite direction (see Fig. 2.33B).

The unshifted radiation (no magnetic field) is at the natural wavelength of the metal resonance line and is absorbed by the Hg atoms of the sample and the molecular absorption. The Zeeman-shifted radiation (Fig. 2.33) (π component) is polarized and is not absorbed by the Hg atoms. However, this radiation is absorbed

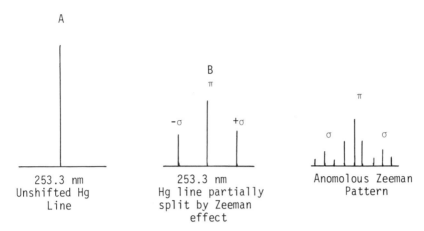

Figure 2.33 Zeeman effect causing shifting of emission lines. π component is at the original wavelength, the ± σ components are shifted up field and down field. In the anomolous pattern there is further splitting. Emission spectrum of Hg 253.3 nm (A) unshifted and (B) after putting the source in a magnetic field.

by the background molecular absorption. The difference between the two absorptions signals is the net atomic absorption signal, which is thereby corrected for the molecular absorption signal.

Experimentally this has been achieved in two ways. First, the applied magnetic field may surround the hollow cathode and be alternating, causing splitting and nonsplitting at its frequency. By tuning the amplifier to this frequency it is possible to discriminate between the split and unsplit radiation. A major difficulty with the technique is that the magnetic field used to generate Zeeman splitting also interacts with the ions in the hollow cathode. This causes the emission from the hollow cathode to be noisy, which in turn introduces imprecision into the procedure (19). A better way is to put the magnetic field around the atomizer. The atoms absorb with no magnet on, but not with the magnet turned on.

A second method is to take advantage of the fact that the split and unsplit light is polarized differently. The magnets may be run A/C or DC. If it is run DC, the π and σ components can be told apart using a polarizing filter. If the field is A/C, a fused silica filter is used. Fused quartz under pressure is birefringent, that is, light is diffracted differently when its axis of polarization is different. The quartz may be stressed at a controlled frequency to yield two axes of polarization and the two beams of light discriminated. By suitable amplification the background absorption—measured by Zeeman-shifted radiation—can be corrected for automatically.

The advantage of this technique is that only one source and only one detector are used. The background correction is made at a wavelength very close to the resonance line and is therefore accurate. The physical paths of the split and unsplit light are identical, as opposed to the use of the D_2 lamp, which may not pass along the identical path as the hollow cathode. Finally, the system has the advantage of being compact and relatively easy to operate.

The complexity of the emission or absorption spectra after Zeeman splitting depends on the magneticc moment of the nucleus and I_1, the spin quantum number. In its simplest form the original line is split into one π and two υ components. With anomolous atoms there may be several π and σ components (20). It is not uncommon for some isotopes of an element to absorb but not others. For this reason, the optimum wavelength for absorption measurements may be changed and the analytical sensitivity may be reduced. Measure values of Zeeman sensitivity ratios (R_2) are shown in Table 2.12.

The Zeeman sensitivity ratio is derived as follows: Beer's law states that

$$A_N = \log \frac{I_0}{I_1}$$

where

A_N = normal absorbance

but

$$A_z = \log \frac{I_H}{I_1}$$

when

A_z = Zeeman absorbance
I_H = measured intensity with magnet on
I_l = measured intensity with magnet off

From this,

$$R_z = \frac{A_z}{A_N}$$

where

R_z = Zeeman sensitivity ratio

The use of a modulated field instead of a fixed field gave a better sensitivity for elements exhibiting anomolous splitting. Also as the magnetic field increased, the splitting of the π component increased causing an increase in R_z. This is in contrast to a fixed magnetic field where increased field strength increases splitting of the υ component resulting in reduced R_z values. These effects are complicated and

Table 2.12 Measured Values of Zeeman Sensitivity Ratios

Element	Wavelength (nm)	Lamp current (mA)	$R_z(\%)$	Element	Wavelength (nm)	Lamp current (mA)	$R_z(\%)$
Ag	328.1	3	80	Nb[a]	334.9	20	80
Al[a]	309.3	5	75	Nd[a]	492.5	10	93
As	193.7	7	78	Ni	232.0	5	75
Au	242.8	4	72	Os[a]	290.9	20	87
B[a]	249.8	15	53	Pb	217.0	5	58
Ba	553.6	10	81		283.3	5	78
Be[a]	234.9	20	48	Pd	244.8	5	78
Bi	223.1	8	62	Pr[a]	495.1	8	85
Ca	422.7	3	84	Pt	265.9	10	70
Cd	228.8	3	77	Rb	780.0	15	86
Co	240.7	5	78	Re[a]	346.0	20	64
Cr	357.9	5	91	Rh	343.5	5	87
Cs	852.1	20	48	Ru	349.9	10	85
Cu	324.8	3	44	Sb	217.6	10	90
	327.4	3	70	Sc[a]	391.2	10	89
Dy[a]	421.2	15	89	Se	196.0	10	74
Er[a]	400.8	10	88	Si[a]	251.6	15	85
Eu[a]	459.4	10	88	Sm[a]	429.7	10	94
Fe	248.3	5	88	Sn[a]	235.5	5	67
Ga[a]	294.4	4	76		286.3	5	77
Gd[a]	368.4	25	88	Sr	460.7	10	80
Ge[a]	265.2	5	87	Ta[a]	271.5	20	75
Hf[a]	307.3	10	92	Tb[a]	432.7	15	88
Hg	253.7	3	67	Te	214.3	8	93
Ho[a]	410.4	15	85	Ti[a]	364.3	20	77
In	303.9	5	39	Tl	276.8	20	69
Ir	208.9	20	88	Tm[a]	371.8	15	80
K	766.5	5	91	U[a]	358.5	20	92
La[a]	550.1	10	82	V[a]	318.5	10	75
Lik	670.8	5	36	W[a]	255.1	20	81
Lu[a]	336.0	10	88	Y[a]	410.2	10	88
Mg	285.2	3	58	Yb[a]	398.8	5	91
Mn	279.5	5	80	Zn	213.9	5	60
Mo	313.3	10	92	Zr[a]	360.1	20	86
Na	589.0	5	88				

[a]Determined in a nitrous oxide-acetylene flame. All other elements determined in air-acetylene.

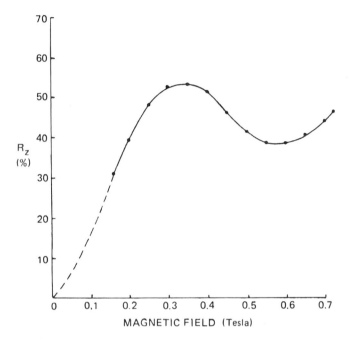

Figure 2.34 Relationship between R_z and magnetic field for copper. (Courtesy of Perkin-Elmer Corp.)

reduced if several isotopes are present or if hyperfine structure develops because of possible overlap of lines (Fig. 2.34).

The R_z value is also dependent on lamp current because this affects self-absorption in the lamp and therefore line shape. Narrower lines give better R_z values (Fig. 2.35).

It has also been established that if $R_z \leq 100\%$ and there is a component of unabsorbed light, then absorbance measurements will reach a maximum and reverse as concentrations increase (reflex) (Fig. 2.36). Reflexing is increased as the magnetic field increases. The problem of reflexing can be serious because two concentrations may give the same absorbance value. Steps must be taken to ensure that this error is eliminated.

The use of the Zeeman effect also affects the noise level of the signal. Photon noise is increased, the effects of lamp flicker decreased, and flame noise decreased. This is because in a sense the Zeeman corrector system acts as a double-beam system in measuring sample and background signals.

d. The Smith-Hiefje Background Corrector

A technique has recently been developed by S. B. Smith, Jr. and G. M. Hiefje (22). This background corrector is based on the fact that the signal from a hollow

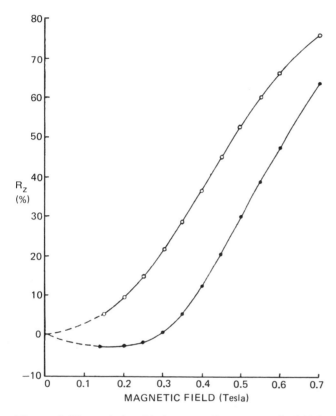

Figure 2.35 Relationship between R_z and magnetic field for zinc at different lamp currents: (O) 3 mA; (●) 6 mA. (Courtesy of Perkin-Elmer Corp.)

cathode varies considerably depending on the current applied to the hollow cathode. If a low current is used (5 milliamps) then a sharp emission line is emitted capable of being absorbed by free atoms in the sample. However, if the hollow cathode is powered by a much higher current, then a much stronger signal is generated but many free atoms are also generated inside a hollow cathode lamp, and they absorb the center of the emitted line. The net result is that the ground-state lines emitted by the hollow cathode under conditions of a high current are strongly reversed and are not absorbed very strongly by the free atoms in the sample (see Fig. 2.5).

The wings of the reversed line, however, are absorbed by the background. Consequently we have the hollow cathode being operated in two modes: (1) with low current, (2) with high current. With low current the narrow line is being

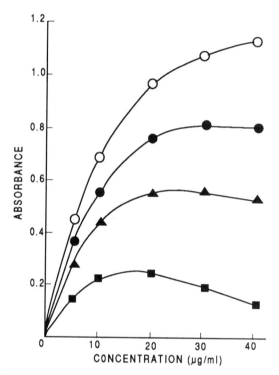

Figure 2.36 Analytical working curves for manganese at different magnetic fields: (O) A_N; (●) A_z at 0.7 T, (▲) A_z at 0.5 T, (■) A_z at 0.3 T. (Courtesy of Perkin-Elmer Corp.)

absorbed by both free atoms and by the background radiation. On the conditions of high current the line is fully absorbed by the free atoms but absorbed just as strongly as by the background.

By using a series of calibration curves the procedure can be used to automatically correct for background. In practice the low current is run for a fairly short period of time and measured. The high current is run for a very short, sharp burst liberating intense emission and free atoms inside the hollow cathode. There is then a delay time to disperse the free atoms before the cycle is started again.

The advantage of the method is that it can be used in single-beam optics; it is not critical to align the beam measuring background absorption and atomic absorption since this is the same beam. This of course is in contrast to the use of the deuterium lamp, which uses different lamps for the two different measurements. In addition the electronics is much simpler than that used in a Zeeman background correction system where polarization is taken care of. An excellent comparison of the methods used for background corrector has been reported by Camrick and Slavin (23).

e. Changes in Sensitivity with Unabsorbed Background Emission Radiation

Even if there is no emitted line from the hollow cathode close to the resonance line, there is always a small amound of background radiation that cannot be resolved from the resonance line. The effect of this unabsorbed radiation is exactly the same as that of an unabsorbed emission line. In general, the intensity is low and no serious error is involved. However, as a matter of principle to avoid any unsuspected problem it is better to operate at as narrow a spectral slit width as is possible. An illustration of the effect of increasing slit width on sensitivity is shown in Table 2.13. There is rapid loss in sensitivity as the mechanical slit width, and hence the spectral slit width, is increased. Under these circumstances, sensitivity is worse and the calibration curve is much flatter than desirable. If it is not possible to resolve two lines that are very close together, the problem sometimes can be overcome if the unabsorbed line originates with the filler gas rather than the metal. In this case a different filler gas such as helium or neon can be used. But if the unabsorbed line is from the element being analyzed, then the only other alternative is to move to a different resonance line which, hopefully, will not be close to an unabsorbable line from the spectrum.

In commercial equipment fixed slits are often used. In this case the manufacturers have built in a fixed mechanical slit width that is considered acceptable for most purposes. For some cases, particularly in research work, a manually adjustable slit is desirable where the slit width can be increased or decreased at will depending on the particular sample being analyzed or studied.

It is always good policy to operate with a maximum of dispersion and a minimum of spectral slit width. But there is no hard and fast rule concerning dispersion and slit width that applies to all elements. The spectrum from each element and filler gas must be examined separately. If there is an unabsorbable

Table 2.13 Effect of Spectral Slit Width on Sensitivity

Mechanical slit width (mm)	Spectral slit width (Å) (Halfwave)	Sensitivity (ppm, 1% absorption)
0.015	3.2 (measured)	5
0.05	3.2 (measured)	5
0.1	4.7 (measured)	6
0.4	7.5 (measured)	7
0.6	11.5 (measured)	10
1.0	19.5 (calculated)	50
1.5	29.5 (calculated)	100

line in the immediate vicinity of the resonance line, then some other precaution must be taken—such as changing the filler gas or moving to another resonance line.

9. Excitation Interference

In flame photometry, when the sample is introduced into the flame, the metal atoms become excited. Suppose that the sample contains two metals: A and B. It is possible that excited metal atoms A transfer energy to unexcited metal atoms B. The intensity of emission from metal B becomes greater. This is called excitation interference and is particularly important with alkaline metal analyses. The effect is important in emission spectrography and flame photometry because it directly affects the number of excited atoms of the elements present in the system. However, the effect is much less important in atomic absorption spectroscopy because it is the unexcited atoms that provide the absorption signal. Excitation interference processes have very little influence on the total number of unexcited atoms present in the system, and therefore interference from this phenomenon is of small importance in atomic absorption spectroscopy.

10. Radiation Interference

In flame photometry the intensity of emission from the sample element is measured at some particular wavelength. If another element present in the sample emits at the same wavelength, there is a direct positive interference called *radiation interference*. The source of this interference is any element or compound that emits at the wavelength being monitored for measurement.

There are four principal sources of radiation interference. These are broad background emission from the flame, background emission from the solvent introduced into the flame, line emission or molecular emission from other elements in the sample, and finally, the emission at the resonance wavelength by the element being determined. Fortunately, this type of interference is not a problem in atomic absorption spectroscopy provided that the equipment is modulated (see page 83).

A problem can arise if a mechanical chopper system is used, in which it is possible for the radiation from the flame to fall on the glass surface of the window of the hollow cathode. Here, it can be reflected back through the chopper system down the light path. Under these conditions it will now be chopped and modulated and will appear to originate in the hollow cathode. The problem can be overcome either by using an electrical chopper system or by slanting the face of the hollow cathode window to prevent reflection down the light path.

If the emission is intense, the photomultiplier detector may become overloaded and noisy. This is particularly likely when highly incandescent flames, such as reducing oxyacetylene flames or flames into which benzene has been introduced, are used. Usually radiation interferences are not a problem with atomic absorption spectroscopy unless intensely incandescent flames are used.

11. Analytical Applications of Atomic Absorption Using Flame Atomizers

The procedures and techniques recommended in this chapter refer to equipment using flame atomizers. Significantly better sensitivity has been achieved using carbon rod and carbon bed atomizers, and these will be considered on page 146.

Atomic absorption spectroscopy has been limited almost exclusively to the determination of metals in liquid samples. Attempts to analyze solid or gas samples have been made and reported in the past. These may work in special cases but have not proven to be very reliable and are not recommended for routine analytical purposes. The normal procedure for analyzing solid and gas samples is to convert them to the liquid state. Solid samples may be dissolved in suitable acid or other solvent. Gas samples may be passed through a liquid scrubbing agent and the element of interest removed from the gas phase trapped in the liquid phase and subsequently analyzed. It should be noted that any pretreatment step (e.g., those necessary to handle solid or gas samples) is a potential source of analytical error. The most common errors are contamination, leading to high answers, and sample loss (e.g., by volatilization during wet ashing) causing low answers. Care should be taken to avoid such problems.

The analytical usefulness of the procedure depends to a great extent on the sensitivity of the process and control of the sources of error.

a. Sensitivity Limits

The *sensitivity limit* is that concentration which absorbs 1% of the signal under the conditions used. This is an artificial definition, but for comparative purposes the universal use of the definition has provided an unambiguous measure for reporting sensitivity data. It avoids problems involved in defining noise level and measuring the analytical signal above the noise level (see S/N ratio, page 75). It avoids problems that arise when detection limits are artificially improved by unrealistically decreasing the noise level with excessive damping. Also, it can be seen that it is independent of the amplification used since both S and N generally are increased. These are all meaningful advantages when comparing data from different laboratories.

The *detection limits* are determined by the noise level and the signal recorded. For some elements the detection limit is significantly better than for other elements recorded under the same definition of sensitivity limits. However, for some elements the reverse is true. This is particularly so when volatile elements such as arsenic and selenium are examined and the pertinent hollow cathodes are noisy.

Most of the data in this chapter refer to sensitivity limits. A summary of the most sensitive sensitivity limits recorded for the various elements in the Periodic Table are shown in Table 2.14. It should be noted that in this table no data are presented for the nonmetallic elements. Iodine (24) has been determined directly, and in-

Table 2.14 Primary Wavelengths and Sensitivities for Atomic Absorption Analysis

Element	Wavelength (nm)	Sensitivity (ppm)	Element	Wavelength (nm)	Sensitivity (ppm)
Al	309.2	1.0	Na	589.0	0.015
Ag	328.1		Nb	405.9	20.0
As	193.7	1.0	Nd	463.4	10.0
Au	242.8		Ni	232.0	0.1
B	249.7	40.0	Os	290.9	1.0
Ba	553.5	0.4	P	213.6	250.0
Be	234.9	0.024	Pb	217.0	0.2
Bi	223.1	0.4	Pd	247.6	0.5
Ca	422.7	0.04	Pr	495.1	10.0
Cd	228.8	0.03	Pt	265.9	1.0
Ce	276.1		Rb	780.0	0.1
Co	240.7	0.15	Re	346.0	15.0
Cr	357.9	0.1	Rh	343.5	0.1
Cs	852.1	0.3	Ru	349.9	0.5
Cu	324.7	0.1	Sb	217.6	0.5
Dy	421.2	0.7	Sc	390.7	0.4
Er	421.2	0.7	Sc	390.7	0.4
Eu	459.4	0.6	Si	251.6	1.0
Fe	248.3		Sm	429.7	8.5
Ga	287.4	2.5	Sn	235.4	0.5
Gd	368.4	15.0	Sr	460.7	0.01
Ge	265.1	2.0	Ta	271.5	10.0
Hf	307.2	1.2	Tb	431.9	5.0
Hg	253.7	1.0	Te	214.3	0.5
Ho	405.4	2.0	Ti	364.3	1.0
In	304.0	0.7	Tl	377.6	0.4
Ir	264.0	8.0	Tm	409.4	0.35
K	766.5	0.04	U	358.5	50.0
La	357.4	45.	V	318.4	1.7
Li	670.7	0.05	W	294.4	1.0
Lu	308.1	6.0	Y	407.7	2.0
Mg	285.2	0.007	Yb	398.8	0.1
Mn	279.5	0.05	Zn	213.9	0.01
Mo	313.3	0.5	Zr	360.1	10.0

The 306.8 nm Bi line is normally considered the most sensitive, but the strength of the adjacent 306.07 nm hydroxyl band head renders it ineffective.

Table 2.15 Characteristics of Selected Elements

Atom	Resonance level (eV)	Emission wavelength (nm)	Ionization potential (eV)	Percent ionized[a]
Na	2.10	589	5.14	99.5
K	1.61	770	4.34	99.5
Mg	4.35	285	7.65	98
Ca	2.93	422	6.11	99
Cu	3.78	327	7.73	90
Zn	5.80	214	9.39	75
N	10.33	120	14.53	0.1
O	9.52	130	13.62	0.1
F	12.98	95	17.42	0.0009
P	6.94	178	10.49	33
S	6.86	181	10.36	14
Cl	8.92	139	12.97	0.9
Br	8.04	163	11.81	5
I	6.77	183	10.45	29

[a]Percent ionized figures based on a temperature of 7500 K and an electron density of $1 \times 10^{15}/cm^3$. Values for Na and K have been recalculated to obtain more significant figures.

direct methods for determination of nonmetallic elements by atomic absorption have been published.

The direct determination of nonmetals by atomic absorption spectroscopy is not practiced mostly because the resonance lines of the nonmetallic elements are in the vacuum ultraviolet (25). Some typical wavelengths are shown in Table 2.15. The sensitivity of halides at various wavelengths are shown in Table 2.16.

The data show that many elements can be determined at concentrations of one part per million or less using flame atomizers. This sensitivity is perfectly satisfactory for the great majority of the elements analyzed commercially.

For some elements where the noise level is very low, scale expanders have been used to expand the 1–10% absorption range to full scale rather than 10% of the recorder scale. This increases the detection limits by approximately 10-fold. In order to use scale expansion it is important that the noise be as low as possible. Most manufacturers of atomic absorption equipment have recommended procedures under which scale expansion can be used with its inherent increase in analytical detectability.

Table 2.16 Comparison of Detection Limits Obtained in VIS and Vacuum-UV Spectral Regions

Element	Wavelength (nm)	Ar-ICP (USN)[a]	Ar-ICP (PN)[b]	He-ICP (USN)	He-MIP (USN)	He-MIP (PN)
					Detection limit (mg/L)	
Cl(I)	134.72	0.008	0.05			
Cl(II)	479.45				0.4	2
Cl(I)	725.67			13		
Br(I)	154.07	0.015	0.05			
Br(II)	478.55				3	60
Br(I)	827.24			5		
I(I)	178.28	0.006	0.02			
I(I)	206.16				0.8	7

[a]PN indicates pneumatic nebulization.
[b]USN indicates ultrasonic nebulization.
Source: Ref. 25.

ABSORPTION WAVELENGTH, PREFERRED FLAMES, AND SENSITIVITIES FOR FLAME ATOMIC ABSORPTION

No data from carbon atomizers have been included in the data in this section.

The author wishes to thank Allied Chemical Company, particularly Dr. Fred Brech, and the Perkin-Elmer Corporation for permission to use their publications on Recommended Procedures. A considerable part of the following information is based on these publications.

Abbreviations Used for Flame Composition

A.A.	air acetylene
N.A.	nitrous oxide acetylene
$O_2A.$	oxygen acetylene
$O_2.H.$	oxygen hydrogen
$O_2.$ air	oxygen air
ox.	oxidizing flame (excess oxidant air, oxygen or nitrous oxide)
red.	reducing flame (excess fuel, acetylene, or hydrogen)

ALUMINUM

Absorption wavelength (nm)	Preferred flame	Sensitivity 1% ab. (ppm)	Analytical range (ppm)
309.3	N.A. red.	1.0	5–50
396.1	N.A. red.	1.3	
308.2	N.A. red.	1.4	
394.4	N.A. red	2.0	
237.3	N.A. red.	3.3	
236.7	N.A. red.	4.0	
257.5	N.A. red.	8.8	

ANTIMONY

Absorption wavelength (nm)	Preferred flame	Sensitivity 1% ab. (ppm)	Analytical range (ppm)
217.6	A.A. (ox.)	0.5	3–40
206.8	A.A. (ox.)	0.7	
231.1	A.A. (ox.)	1.2	

ARSENIC

Absorption wavelength (nm)	Preferred flame	Sensitivity 1% ab. (ppm)	Analytical range (ppm)
189.0	A. H. (ox.)	1	5–50
193.7	A. H. (ox.)	2	50
197.2	A. H. (ox.)	3	

BARIUM

Absorption wavelength (nm)	Preferred flame	Sensitivity 1% ab. (ppm)	Analytical range (ppm)
553.5	N. A. (red.)	0.4	3–25
350.1	N. A. (red.)	5.0	

BERYLLIUM

Absorption wavelength (nm)	Preferred flame	Sensitivity 1% ab. (ppm)	Analytical range (ppm)
234.8	N. A. (red.)	0.02	0.2–4

BISMUTH

Absorption wavelength (nm)	Preferred flame	Sensitivity 1% ab. (ppm)	Analytical range (ppm)
223.1	A. A. (ox.)	0.4	2–50
222.8	A. A. (ox.)	1.5	
306.7	A. A. (ox.)	2.1	
206.2	A. A. (ox.)	5.5	

BORON

Absorption wavelength (nm)	Preferred flame	Sensitivity 1% ab. (ppm)	Analytical range (ppm)
249.7	N. A. (red.)	40	100–900
249.6	N. A. (red.)	100	

CADMIUM

Absorption wavelength (nm)	Preferred flame	Sensitivity 1% ab. (ppm)	Analytical range (ppm)
228.8	A. A. (ox.)	0.03	0.2–2.0
326.1	A. A. (ox.)	1.0	

CALCIUM

Absorption wavelength (nm)	Preferred flame	Sensitivity 1% ab. (ppm)	Analytical range (ppm)
422.7	N. A. (red.)	0.04	0.2–7
239.8	N. A. (red.)	20	

CERIUM

Absorption wavelength (nm)	Preferred flame	Sensitivity 1% ab. (ppm)	Analytical range (ppm)
520.0	N. A.	30	

CESIUM

Absorption wavelength (nm)	Preferred flame	Sensitivity 1% ab. (ppm)	Analytical range (ppm)
852.1	A. Coal Gas or A. A. (ox.)	0.2	2.0–15
894.3	A. Coal Gas or A. A. (ox.)	0.2	
455.6	A. Coal Gas or A. A. (ox.)	20.0	
459.3	A. Coal Gas or A. A. (ox.)		

CHROMIUM

Absorption wavelength (nm)	Preferred flame	Sensitivity 1% ab. (ppm)	Analytical range (ppm)
357.9	A. A. (red.)	0.1	0.5–5
359.3	A. A. (red.)	0.1	
425.4	A. A. (red.)	0.2	

COBALT

Absorption wavelength (nm)	Preferred flame	Sensitivity 1% ab. (ppm)	Analytical range (ppm)
240.7	A. A. (ox.)	0.15	0.6–5
242.5	A. A. (ox.)	.20	
252.1	A. A. (ox.)	0.3	
241.1	A. A. (ox.)	0.3	
352.7	A. A. (ox.)	2.2	
347.4	A. A. (ox.)	7.5	

COPPER

Absorption wavelength (nm)	Preferred flame	Sensitivity 1% ab. (ppm)	Analytical range (ppm)
324.7	A. A. (ox.)	0.1	0.5–10
327.4	A. A. (ox.)	0.2	
217.8	A. A. (ox.)	0.4	
216.5	A. A. (ox.)	0.7	
222.6	A. A. (ox.)	1.5	
249.2	A. A. (ox.)	7.0	
224.4	A. A. (ox.)	16.0	
244.2	A. A. (ox.)	30.0	

DYSPROSIUM

Absorption wavelength (nm)	Preferred flame	Sensitivity 1% ab. (ppm)	Analytical range (ppm)
421.2	N. A. (red.)	0.9	4.0–2.0
404.6	N. A. (red.)	0.8	
418.7	N. A. (red.)	0.9	
419.5	N. A. (red.)	1.0	

ERBIUM

Absorption wavelength (nm)	Preferred flame	Sensitivity 1% ab. (ppm)	Analytical range (ppm)
40.8	N. A. (red.)	1.0	5.0–40
415.1	N. A. (red.)	2.3	
386.3	N. A. (red.)	2.0	
389.3	N. A. (red.)	2.4	
408.8	N. A. (red.)	6.0	
393.7	N. A. (red.)	6.5	
381.0	N. A. (red.)	7.0	
390.5	N. A. (red.)	20.0	
394.4	N. A. (red.)	21.0	
460.7	N. A. (red.)	22.0	

EUROPIUM

Absorption wavelength (nm)	Preferred flame	Sensitivity 1% ab. (ppm)	Analytical range (ppm)
459.4	N. A. (red.)	0.6	3.0–50
462.7	N. A. (red.)	2.0	
466.2	N. A. (red.)	2.0	
322.1	N. A. (red.)	7.0	
321.3	N. A. (red.)	9.0	
311.1	N. A. (red.)	9.0	
333.4	N. A. (red.)	12.0	

GADOLINIUM

Absorption wavelength (nm)	Preferred flame	Sensitivity 1% ab. (ppm)	Analytical range (ppm)
407.9	N. A. (red.)	20	100–1000
368.4	N. A. (red.)	25	
378.3	N. A. (red.)	25	
405.8	N. A. (red.)	25	
405.4	N. A. (red.)	25	
371.4	N. A. (red.)	30	
419.4	N. A. (red.)	40	
367.4	N. A. (red.)	43	
404.5	N. A. (red.)	50	
394.6	N. A. (red.)	100	

GALLIUM

Absorption wavelength (nm)	Preferred flame	Sensitivity 1% ab. (ppm)	Analytical range (ppm)
287.4	A. A. (ox.)	2.5	15.0–200
294.4	A. A. (ox.)	2.5	
417.2	A. A. (ox.)	4.0	
250.0	A. A. (ox.)	20.0	
245.0	A. A. (ox.)	20.0	
272.0	A. A. (ox.)	50.0	

GERMANIUM

Absorption wavelength (nm)	Preferred flame	Sensitivity 1% ab. (ppm)	Analytical range (ppm)
265.1	N. A. (red.)	2.0	20–200
265.2	N. A. (red.)	2.0	
259.2	N. A. (red.)	5.0	
271.0	N. A. (red.)	6.0	
275.5	N. A. (red.)	6.5	
269.1	N. A. (red.)	9.0	

GOLD

Absorption wavelength (nm)	Preferred flame	Sensitivity 1% ab. (ppm)	Analytical range (ppm)
242.8	A. A. (red.)	0.3	2.5–20
267.6	A. A. (red.)	0.4	
274.8	A. A. (red.)	250.0	
312.8	A. A. (red.)	240.0	

HAFNIUM

Absorption wavelength (nm)	Preferred flame	Sensitivity 1% ab. (ppm)	Analytical range (ppm)
307.3	N. A. (red.)	15	100–500
286.6	N. A. (red.)	15	
289.8	O_2. A.	70	
296.5	O_2. A.	80	
368.2	O_2. A.	80	

HOLMIUM

Absorption wavelength (nm)	Preferred flame	Sensitivity 1% ab. (ppm)	Analytical range (ppm)
410.4	N. A.	2.0	
416.3	N. A.	3.0	
405.4	N. A.	3.0	

INDIUM

Absorption wavelength (nm)	Preferred flame	Sensitivity 1% ab. (ppm)	Analytical range (ppm)
304.9	A. A.	0.7	
325.6	A. A.	0.7	
410.5	A. A.	2.0	
275.4	A. A.	20.0	

IRIDIUM

Absorption wavelength (nm)	Preferred flame	Sensitivity 1% ab. (ppm)	Analytical range (ppm)
208.9	A. A. (red.)	10.0	50–1000
264.0	A. A. (red.)	10.0	
766.5	A. A. (red.)	8.0	
769.9	A. A. (red.)	19.0	
404.4	A. H.	4000.0	

IRON

Absorption wavelength (nm)	Preferred flame	Sensitivity 1% ab. (ppm)	Analytical range (ppm)
248.3	A. A.	0.1	1.5
248.8	A. A.	0.2	
252.3	A. A.	0.2	
271.9	A. A.	0.3	
302.1	A. A.	0.4	
250.1	A. A.	0.4	
216.7	A. A.	0.5	
372.1	A. A.	0.7	
296.7	A. A.	1.0	
386.0	A. A.	2.0	
344.1	A. A.	2.5	
368.0	A. A.	10.0	

LANTHANUM

Absorption wavelength (nm)	Preferred flame	Sensitivity 1% ab. (ppm)	Analytical range (ppm)
550.1	A. A. (ox.)	35	200–2500
418.7	A. A. (ox.)	50	
495.0	A. A. (ox.)	50	
357.4	A. A. (ox.)	110	
365.0	A. A. (ox.)	150	
392.8	A. A. (ox.)	150	

LEAD

Absorption wavelength (nm)	Preferred flame	Sensitivity 1% ab. (ppm)	Analytical range (ppm)
217.0	A. A. (ox.)	0.2	1–10
283.3	A. A. (ox.)	0.5	2–20
261.4	A. A. (ox.)	5.0	
368.4	A. A. (ox.)	12.0	

LITHIUM

Absorption wavelength (nm)	Preferred flame	Sensitivity 1% ab. (ppm)	Analytical range (ppm)
670.8	A. A. (ox.)	0.03	0.1–3.0
323.3	A. A. (ox.)	10	
610.4	A. A. (ox.)	100.0	

LUTETIUM

Absorption wavelength (nm)	Preferred flame	Sensitivity 1% ab. (ppm)	Analytical range (ppm)
336.0	N. A. (red.)	6.0	50–500
331.2	N. A. (red.)	11.0	
337.7	N. A. (red.)	12.0	
356.0	N. A. (red.)	13.0	
398.9	N. A. (red.)	55.0	
451.9	N. A. (red.)	66.0	

MAGNESIUM

Absorption wavelength (nm)	Preferred flame	Sensitivity 1% ab. (ppm)	Analytical range (ppm)
285.2	A. A. (ox.)	0.007	0.01–0.5
202.6	A. A. (ox.)	0.2	
279.6	A. A. (ox.)	5.0	

MANGANESE

Absorption wavelength (nm)	Preferred flame	Sensitivity 1% ab. (ppm)	Analytical range (ppm)
279.5	A. H. (ox.)	0.05	0.5–5.0
280.1	A. H. (ox.)	0.08	.
403.1	A. H. (ox.)	0.5	
279.2	A. H. (ox.)	50.0	

MERCURY

Absorption wavelength (nm)	Preferred flame	Sensitivity 1% ab. (ppm)	Analytical range (ppm)
185.0	N. A. (red.)	0.5	10–300
253.7	N. A. (red.)	1.0	

MOLYBDENUM

Absorption wavelength (nm)	Preferred flame	Sensitivity 1% ab. (ppm)	Analytical range (ppm)
313.3	N. A. (red.)	0.5	5.0–60.0
317.0	N. A. (red.)	0.8	
379.8	N. A. (red.)	1.0	
319.4	N. A. (red.)	1.0	
386.4	N. A. (red.)	1.2	
390.3	N. A. (red.)	1.6	
315.8	N. A. (red.)	2.0	
320.9	N. A. (red.)	4.3	
311.2	N. A. (red.)	10.0	

NEODYMIUM

Absorption wavelength (nm)	Preferred flame	Sensitivity 1% ab. (ppm)	Analytical range (ppm)
334.9	N. A. (ox.)	10	10–700
463.4	N. A. (ox.)	10	
492.4	N. A. (ox.)	14	
471.9	N. A. (ox.)	20	

NICKEL

Absorption wavelength (nm)	Preferred flame	Sensitivity 1% ab. (ppm)	Analytical range (ppm)
232.0	N. A. (red.)	0.1	1.0–1000
231.1	N. A. (red.)	0.2	
352.5	N. A. (red.)	0.3	
341.5	N. A. (red.)	0.4	
304.1	N. A. (red.)	0.4	
341.2	N. A. (red.)	0.7	
351.5	N. A. (red.)	0.8	
303.8	N. A. (red.)	1.2	
337.0	N. A. (red.)	1.8	
323.0	N. A. (red.)	3.0	
294.4	N. A. (red.)	5.5	

NIOBIUM

Absorption wavelength (nm)	Preferred flame	Sensitivity 1% ab. (ppm)	Analytical range (ppm)
334.4	N. A. (red.)	20	100–1000
358.0	N. A. (red.)	22	
334.9	N. A. (red.)	24	
408.9	N. A. (red.)	28	
335.0	N. A. (red.)	30	
412.4	N. A. (red.)	38	
357.6	N. A. (red.)	50	
353.5	N. A. (red.)	60	
374.0	N. A. (red.)	64	
415.3	N. A. (red.)	100	

OSMIUM

Absorption wavelength (nm)	Preferred flame	Sensitivity 1% ab. (ppm)	Analytical range (ppm)
290.9	N. A. (red.)	1.0	10–200
305.9	N. A. (red.)	1.6	
263.7	N. A. (red.)	1.8	
301.8	N. A. (red.)	3.0	
330.2	N. A. (red.)	3.5	
271.5	N. A. (red.)	4.0	
280.7	N. A. (red.)	4.5	
264.4	N. A. (red.)	5.0	
442.0	N. A. (red.)	20.0	
426.1	N. A. (red.)	30.0	

PALLADIUM

Absorption wavelength (nm)	Preferred flame	Sensitivity 1% ab. (ppm)	Analytical range (ppm)
244.8	A. A. (red.)	0.3	5.0–50.0
247.6	A. A. (red.)	0.2	
276.3	A. A. (red.)	1.3	
340.5	A. A. (red.)	1.5	

PHOSPHORUS

Absorption wavelength (nm)	Preferred flame	Sensitivity 1% ab. (ppm)	Analytical range (ppm)
178.3	N. A. (red.)	5.0	2000–10,000
213.6	N. A. (red.)	250	
213.5	N. A. (red.)	250	
214.9	N. A. (red.)	500	

PLATINUM

Absorption wavelength (nm)	Preferred flame	Sensitivity 1% ab. (ppm)	Analytical range (ppm)
265.9	A. A. (ox.)	1.0	5–100
306.5	A. A. (ox.)	2.0	
283.0	A. A. (ox.)	3.5	
293.0	A. A. (ox.)	3.8	
273.4	A. A. (ox.)	4.0	
217.5	A. A. (ox.)	4.0	
248.7	A. A. (ox.)	5.0	
299.8	A. A. (ox.)	5.5	
271.9	A. A. (ox.)	9.0	

POTASSIUM

Absorption wavelength (nm)	Preferred flame	Sensitivity 1% ab. (ppm)	Analytical range (ppm)
766.5	A. A. (red.)	0.03	0.2–2.0
769.9	A. A. (red.)	0.05	
404.4	A. A. (red.)	10.0	
404.7	A. A. (red.)	10.0	

PRASEODYMIUM

Absorption wavelength (nm)	Preferred flame	Sensitivity 1% ab. (ppm)	Analytical range (ppm)
495.1	N. A.	15.0	

RHENIUM

Absorption wavelength (nm)	Preferred flame	Sensitivity 1% ab. (ppm)	Analytical range (ppm)
346.0	N. A. (red.)	15	100–1000
346.5	N. A. (red.)	25	
345.2	N. A. (red.)	35	

RHODIUM

Absorption wavelength (nm)	Preferred flame	Sensitivity 1% ab. (ppm)	Analytical range (ppm)
343.5	A. A. (ox.)	0.4	0.5–20
369.2	A. A. (ox.)	0.5	
339.7	A. A. (ox.)	0.7	
350.2	A. A. (ox.)	1.0	
365.8	A. A. (ox.)	1.8	
370.1	A. A. (ox.)	3.0	
350.7	A. A. (ox.)	10.0	

RUBIDIUM

Absorption wavelength (nm)	Preferred flame	Sensitivity 1% ab. (ppm)	Analytical range (ppm)
780.0	A. A. (ox.)	0.1	0.5–7.0
794.8	A. A. (ox.)	0.2	
420.2	A. A. (ox.)	12.0	
421.6	A. A. (ox.)	25.0	

RUTHENIUM

Absorption wavelength (nm)	Preferred flame	Sensitivity 1% ab. (ppm)	Analytical range (ppm)
349.9	A. A. (ox.)	0.5	3–50
372.8	A. A. (ox.)	0.8	
379.9	A. A. (ox.)	1.1	
392.6	A. A. (ox.)	6.0	

SAMARIUM

Absorption wavelength (nm)	Preferred flame	Sensitivity 1% ab. (ppm)	Analytical range (ppm)
429.7	N. A. (red.)	8.0	50.0–500.0
476.0	N. A. (red.)	20	
511.7	N. A. (red.)	20	
520.1	N. A. (red.)	25	
476.0	N. A. (red.)	40	
458.3	N. A. (red.)	50	

SCANDIUM

Absorption wavelength (nm)	Preferred flame	Sensitivity 1% ab. (ppm)	Analytical range (ppm)
391.2	N. A. (red.)	0.4	2.0–25.0
390.8	N. A. (red.)	0.4	
402.4	N. A. (red.)	1.0	
405.5	N. A. (red.)	1.1	
327.0	N. A. (red.)	2.0	
408.2	N. A. (red.)	3.0	
327.4	N. A. (red.)	5.0	

SELENIUM

Absorption wavelength (nm)	Preferred flame	Sensitivity 1% ab. (ppm)	Analytical range (ppm)
196.0	A. A. (ox.)	0.5	2.0–50.0
204.0	A. A. (ox.)	1.5	
206.3	A. A. (ox.)	6.0	
207.5	A. A. (ox.)	20.0	

SILICON

Absorption wavelength (nm)	Preferred flame	Sensitivity 1% ab. (ppm)	Analytical range (ppm)
251.6	N. A. (red.)	1.0	5.0–100
250.7	N. A. (red.)	3.0	
252.8	N. A. (red.)	3.5	
252.4	N. A. (red.)	4.0	
221.7	N. A. (red.)	4.5	
221.1	N. A. (red.)	8.0	

SILVER

Absorption wavelength (nm)	Preferred flame	Sensitivity 1% ab. (ppm)	Analytical range (ppm)
328.1	A. A. (ox.)	0.1	1.0–20
338.3	A. A. (ox.)	0.15	

SODIUM

Absorption wavelength (nm)	Preferred flame	Sensitivity 1% ab. (ppm)	Analytical range (ppm)
589.0	A. A.	0.02	0.1–1.0
	or A. H.		
589.5	A. A.	0.02	
	or A. H.		
330.2	A. A.	0.30	
	or A. H.		
330.3	A. A.	0.30	
	or A. H.		

STRONTIUM

Absorption wavelength (nm)	Preferred flame	Sensitivity 1% ab. (ppm)	Analytical range (ppm)
460.7	A. A. (red.)	0.1	1.0–10.0
	or A. H.		
407.8	A. A. (red.)	7.0	
	or A. H.		

SULFUR

Absorption wavelength (nm)	Preferred flame	Sensitivity 1% ab. (ppm)	Analytical range (ppm)
180.7	A. A.	9.0	No Data

TANTALUM

Absorption wavelength (nm)	Preferred flame	Sensitivity 1% ab. (ppm)	Analytical range (ppm)
271.5	N. A. (red.)	10.0	600–1200
260.8	N. A. (red.)	20.0	
260.9	N. A. (red.)	20.0	
264.7	N. A. (red.)	40	
293.5	N. A. (red.)	40	
255.9	N. A. (red.)	40	
265.3	N. A. (red.)	50	
269.8	N. A. (red.)	50	
275.8	N. A. (red.)	80	

TECHNETIUM

Absorption wavelength (nm)	Preferred flame	Sensitivity 1% ab. (ppm)	Analytical range (ppm)
261.4	A. A. (red.)	3.0	10–70
261.6	A. A. (red.)	3.0	
260.9	A. A. (red.)	12.0	
429.7	A. A. (red.)	20.0	
426.2	A. A. (red.)	25.0	
318.2	A. A. (red.)	30.0	
423.8	A. A. (red.)	33	
363.6	A. A. (red.)	33	
317.3	A. A. (red.)	300	
346.6	A. A. (red.)	300	
403.2	A. A. (red.)	300	

TELLURIUM

Absorption wavelength (nm)	Preferred flame	Sensitivity 1% ab. (ppm)	Analytical range (ppm)
214.3	A. A. or A. H.	0.5	2.0–23
225.9	A. A. or A. H.	4.0	
238.6	A. A. or A. H.	70.0	

TERBIUM

Absorption wavelength (nm)	Preferred flame	Sensitivity 1% ab. (ppm)	Analytical range (ppm)
432.6	N. A. (red.)	5.0	40–600
431.9	N. A. (red.)	6.0	
390.1	N. A. (red.)	8.0	
406.2	N. A. (red)	9.0	
433.8	N. A. (red.)	10.0	
410.5	N. A. (red.)	20.0	

THALLIUM

Absorption wavelength (nm)	Preferred flame	Sensitivity 1% ab. (ppm)	Analytical range (ppm)
276.8	A. A. (ox.)	0.4	2.0–20.0
377.6	A. A. (ox.)	2.0	
238.0	A. A. (ox.)	3.0	
258.0	A. A. (ox.)	10.0	

THORIUM

Absorption wavelength (nm)	Preferred flame	Sensitivity 1% ab. (ppm)	Analytical range (ppm)
324.6	N. A.	500	

THULIUM

Absorption wavelength (nm)	Preferred flame	Sensitivity 1% ab. (ppm)	Analytical range (ppm)
371.8	N. A. (red.)	0.4	3.0–60.0
410.6	N. A. (red.)	5.0	
374.4	N. A. (red.)	5.0	
409.4	N. A. (red.)	6.0	
418.8	N. A. (red.)	6.0	
420.4	N. A. (red.)	15.0	
375.2	N. A. (red.)	20.0	
436.0	N. A. (red.)	25.0	
341.0	N. A. (red.)	30.0	

TIN

Absorption wavelength (nm)	Preferred flame	Sensitivity 1% ab. (ppm)	Analytical range (ppm)
224.6	A. A. (red.)	0.2	5.0–100
286.3	A. A. (red.)	1.0	
235.5	A. A. (red.)	1.0	
270.6	A. A. (red.)	2.0	
303.4	A. A. (red.)	2.0	
253.5	A. A. (red.)	3.0	
219.9	A. A. (red.)	4.0	
300.9	A. A. (red.)	5.0	
235.5	A. A. (red.)	5.0	
266.1	A. A. (red.)	15.0	

TITANIUM

Absorption wavelength (nm)	Preferred flame	Sensitivity 1% ab. (ppm)	Analytical range (ppm)
365.3	N. A. (red.)	1.0	6.0–150
364.3	N. A. (red.)	1.1	
320.0	N. A. (red.)	1.2	
363.6	N. A. (red.)	1.2	
335.5	N. A. (red.)	1.5	
375.3	N. A. (red.)	1.7	
337.2	N. A. (red.)	1.7	
399.9	N. A. (red.)	1.7	
399.0	N. A. (red.)	2.0	

TUNGSTEN

Absorption wavelength (nm)	Preferred flame	Sensitivity 1% ab. (ppm)	Analytical range (ppm)
400.9	N. A. (red.)	5.0	10–500
255.1	N. A. (red.)	5.0	
294.4	N. A. (red.)	10.0	
268.1	N. A. (red.)	10.0	
272.4	N. A. (red.)	10.0	
294.7	N. A. (red.)	10.0	
283.1	N. A. (red.)	20.0	
289.6	N. A. (red.)	25.0	
287.9	N. A. (red.)	35.0	
430.2	N. A. (red.)	80.0	

URANIUM

Absorption wavelength (nm)	Preferred flame	Sensitivity 1% ab. (ppm)	Analytical range (ppm)
351.5	N. A. (red.)	50	200–2000
358.5	N. A. (red.)	50	
356.7	N. A. (red.)	80	

VANADIUM

Absorption wavelength (nm)	Preferred flame	Sensitivity 1% ab. (ppm)	Analytical range (ppm)
318.4	N. A. (red.)	1.0	5–150
306.6	N. A. (red.)	3.0	
306.0	N. A. (red.)	3.0	
305.6	N. A. (red.)	5.0	
320.2	N. A. (red.)	10.0	
390.2	N. A. (red.)	10.0	

YTTERBIUM

Absorption wavelength (nm)	Preferred flame	Sensitivity 1% ab. (ppm)	Analytical range (ppm)
398.8	N. A. (red.)	0.7–7.0	
346.4	N. A. (red.)		
246.4	N. A. (red.)		
267.2	N. A. (red.)		

YTTRIUM

Absorption wavelength (nm)	Preferred flame	Sensitivity 1% ab. (ppm)	Analytical range (ppm)
410.2	N. A. (red.)	2.0	20–200
407.7	N. A. (red.)	5.0	
412.8	N. A. (red.)	5.0	
414.3	N. A. (red.)	2.8	
362.1	N. A. (red.)	4.0	

ZINC

Absorption wavelength (nm)	Preferred flame	Sensitivity 1% ab. (ppm)	Analytical range (ppm)
213.9	N. A. (red.)	0.01	0.1–1.2
307.6	N. A. (red.)	50.0	

ZIRCONIUM

Absorption wavelength (nm)	Preferred flame	Sensitivity 1% ab. (ppm)	Analytical range (ppm)
360.1	N. A. (red.)	10.0	50–600
354.8	N. A. (red.)	15.0	
303.0	N. A. (red.)	15	
301.2	N. A. (red.)	17	
248.2	N. A. (red.)	18	
362.4	N. A. (red.)	20	

b. Qualitative Analysis

Inasmuch as only one element (sometimes up to three) is usually measured at any one time, atomic absorption spectroscopy is not suitable for qualitative analysis unless specific elements are being tested for. For a sample of unknown composition, other techniques such as X-ray fluorescence, plasma mass spectrometry, or optical emission spectrography are usually much more useful.

c. Quantitative Analysis

Quantitative measurement is one of the ultimate objectives of analytical chemistry. Atomic absorption is an excellent quantitative method. It is deceptively easy to use, particularly when flame atomizers are utilized. Calibration methods and the reliability of results is treated on page 30.

i. Quantitative Analytical Range Like all other spectroscopic quantitative analytical procedures, there is a maximum and a minimum to the concentration range of application. The minimum of the range is a function of the detection limits of the element under optimum conditions. The ultimate limiting factor is the noise level of the instrument being used. The maximum of the analytical range is determined by the degree of absorption by the sample. At high concentrations the degree of absorption is very high. Small changes in concentration of the sample produce virtually no changes in absorption by the sample. Hence, it is difficult or impossible to measure absorption changes caused by concentration changes in the sample. (See Ringbom plot, page 40).

ii. Extension of Maximum Absorption Range   In practice, the upper concentration limit can be extended by three methods. The first method is simply to dilute the sample until the concentration of the element being determined is reduced to an acceptable analytical range. The second method can be used when elongated burners are employed., By placing the burner perpendicular to the light path rather than along the light path, much of the flame is removed from the opticcal system. The number of absorbing atoms in the light is decreased and there is reduction in sensitivity. In practice this method is unwieldy, but it can be used when necessary. One problem with this technique is that the calibration curves used for quantitative analysis must be prepared under exactly the same conditions as those under which measurements are taken.

The third method is to use an absorption line with a lower oscillator strength. The degree of absorption is directly related to the oscillator strength of the resonance line used, as can be seen from Eq. (1.7). If an element has two resonance lines, one of which has an oscillator strength 10 times as great as the second line, then the calibration curves will cover dynamic ranges, one of which is 10 times as great as the other (see Fig. 2.33). It is by far the most convenient method of extending the analytical range, because it can be achieved by merely changing the wavelength of the monochromator. No other changes in optical alignment or sample treatment are involved. One difficulty with the method is that not all elements have several resonance lines that can be used at will. In practice, if a solution contains a very high percentage of an element to be determined, then any one of the above procedures can be used and, if necessary, all three.

D. RECOMMENDED PROCEDURES FOR QUANTITATIVE ANALYSIS

Analytical procedures and recommended conditions for obtaining quantitative reliable results are provided by the major atomic absorption instrumentation manufacturers. These conditions vary somewhat from one manufacturer to another, but in general there is a significant degree of compatibility among the techniques recommended. The following is a synopsis of conditions recommended for use with flame atomizers. Using these conditions, reproducible quantitative results have been obtained on a routine basis.

Several types of flames have found routine use, including oxyhydrogen, oxy-acetylene, air-acetylene, and nitrous oxide-acetylene flames.

The sensitivity is defined as that concentration of solution that will lead to 1% absorption when atomized in a flame. The detection limit is that concentration that provides a signal that is analyticcally detectable. It is based on the signal-to-noise ratio. The detection limit necessarily depends very much on instrumental conditions—particularly signal damping, which provides a smoother (less noisy) signal. Unfortunately, it is also necessary to take a longer period of time to come to a steady signal when the signal is damped. Care must be taken in interpreting detection limits, because frequently the conditions are idealized and may not be easily achieved in routine operation.

1. Calibration Curves

a. Preparation of Standard Solutions

It is most important that the solvent used in preparing the standard solution be the same as the solvent of the sample. This will be readily understood if we remember the profound effects that the solvent and the matrix have on the absorption signal for a given metal concentration.

It is also important to remember that when very dilute solutions are prepared for calibration purposes, the metal can plate out onto the walls of the container within a relatively short period of time (e.g., one or two hours). The plating out occurs along strain lines in the glass and is not reproducible from one bottle to another. It is particularly important if the concentration is less than one part per million. To avoid error caused by plating out of the sample element, it is necessary to prepare standard solutions and run calibration data very shortly thereafter. This problem is reduced by the use of Teflon containers. Suitable salts for preparing calibration curves are listed in Table 2.17.

To a first approximation, absorption by atoms follows Beer's law. This equation indicates that there is a linear relationship between absorbance and concentration provided the burner path length is kept constant. This relationship holds only cner ideal conditions; it is seldom realized in practice. As concentration increases, the relationship usually deviates from linearity so that at higher concentrations there is a reduced increase in absorbance per unit concentration. This deviation can be accommodated by preparing suitable calibration curves and corrected automatically in commercial equipment.

b. Standard Calibration Curves

Calibration curves are determined experimentally by preparing a series of solutions, each with a known concentration of the absorbing solutions. The absorbance of each solution is then meassured and a curve is prepared relating the experimentally determined absorbance and the concentration of the solution. The typical calibration curve in Fig. 2.37 was prepared by plotting the absorbance against the concentration of the solution.

When a solution of unknown concentration is to be determined, it is nebulized in the flame spectrometer and I_1 (or A) is measured. Using the calibration curve, we ascertain the experimental relationship between I_1 and c. By measuring I_1 when the sample is absorbing radiation, we can determine the concentration of the absorbing compound in the sample.

If the absorbance by an "unknown" sample were 0.5, the calibration curve would show that the concentration was 2.3 ppm.

Table 2.17 Salts Suitable for Preparing Calibration Curves

Metal	Salt
Aluminum	$AlCl_3$ (from metal) in approx. 2.25 N HCl
Arsenic	As_2O_3 dissolved in 20% NaOH; neutralize with dilute H_2SO_4
Barium	$BaCl_2$ in water
Bismuth	$Bi(NO_3)_3$ (from metal) in dilute HNO_3
Boron	H_3BO_3 in water
Cadmium	$CdCl_2$ (from metal) in approx. 4.5 N HCl
Calcium	$CaCl_2$ (from $CaCO_3$) in dilute HCl
Cesium	CsCl in water
Chromium	K_2CrO_4 in water
Copper	$Cu(NO_3)_2$ (from metal) in dilute HNO_3
Iron	$Fe(NO_3)_3$ (from metal) in 0.2 N HNO_3
Lead	$Pb(NO_3)_2$ in 1% HNO_3
Lithium	LiCl (from Li_2CO_3) in dilute HCl
Magnesium	$MgCl_2$ (from metal) in dilute HCl
Mercury	$HgCl_2$ in 0.25 N H_2SO_4
Molybdenum	$(NH_4)_2MoO_4$ in 10% NH_4OH
Nickel	$Ni(NO_3)_2$ (from metal) in dilute HNO_3
Potassium	KCl in water
Rubidium	RbCl in water
Silicon	Na_2SiO_3 (from SiO_2) in water
Silver	$AgNo_3$ in water
Sodium	NaCl in water
Strontium	$Sr(NO_3)_2$ in water
Tungsten	$Na_2WO_4 \cdot 2H_2O$ in water (1% NaOH added)
Zinc	$ZnCl_2$ (from metal) in dilute HCl
Zirconium	$ZrOCl_2 \cdot 8H_2O$ in 20% HCl

As mentioned previously, concentration of solutions, particularly very dilute solutions, may change with time. If 1% accuracy is required, it is good practice to prepare standard solutions daily from stock solutions of 500 or 1000 µg/ml. It has been found that the presence of 1% KNO_3 retards the plating of the metal to the sides of the container and lengthens the life of the standard.

c. Standard Addition Method

The standard addition method may be used if no suitable calibration curves have been prepared and it is not convenient to prepare such a curve, for example, because of the time delay in preparing a calibration curve or lack of information on the solvent and matrix in the sample. In a typical example, the concentration of sodium in a plant stream of unknown composition may be determined by the

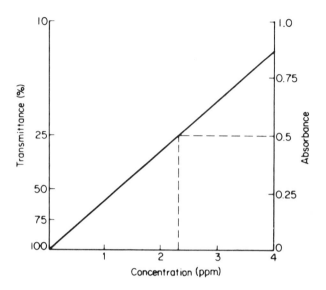

Figure 2.37   Calibration curve for zinc at 213.8 nm.

method of standard addition. Several aliquots, each of 100 ml of the sample, are taken, and different, but known, quantities of the metal being determined are added to each aliquot except one, which is left untreated. The absorption by each aliquot is then measured. The results are shown in Table 2.18.

The data can be displayed in the form of a graph (Fig. 2.37). A measure of the background absorption can be made using a similar sample containing none of the elements being determined or by measuring the absorption at a wavelength slightly different from the characteristic wavelength of the metal. It can be seen from Fig. 2.38 that, over the range examined, the relationship between absorption and sodium concentration added was linear. From the difference in readings between samples 1 and 2, 2 and 3, and 3 and 4, we can calculate the increased absorbance due to 1 ppm Na. From the total absorbance by the sample, the apparent Na concentration can be calculated.

This relationship can be summarized as:

Sample concentration = absorption by sample × absorption per unit concentration

= total absorption − background absorption x

$$= (A - B) \ \frac{\delta \text{Absorption}}{\delta \text{Concentration}}$$

Table 2.18

Sample	Absorbance	Na added to sample (ppm)
1	0.2	0
2	0.42	1
3	0.55	2
4	0.68	3
5[a]	0.05	Background absorption by flame

[a]Flame only; no sample added.

where

A = absorption by the sample

B = background absorption

$\dfrac{\delta \text{Absorption}}{\delta \text{Concentration}}$ = rate of change of absorption with change in sample concentration.

Similarly, data can be obtained by projecting the Na concentration axis in a negative direction (to the left) and measuring the point of concentration at the intersection of this line and the absorption line. This is shown in Fig. 2.38B.

Correction for background is performed as illustrated above. The absorption measurement is put onto the vertical and a line parallel to the sodium A/C line drawn. The point of intersection with the horizontal indicates the concentration error equivalent to the background error. In Fig. 2.38C, the background level to an equivalent concentration of 0.9, subtracting this from the total of 2.9 ppm, led to a corrected concentration of (2.9 − 0.9) = 2.0 ppm.

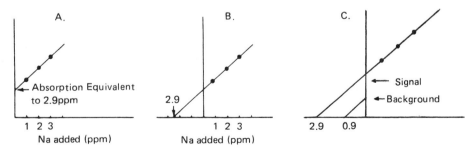

Figure 2.38 Determination of sodium by the standard addition method. Notice background correction in case C.

E. ELECTROTHERMAL ATOMIZATION

The commercial carbon atomizers currently available have been a triumph for physics and a disaster for analytical chemistry. It has been a triumph to get such an unlikely and unreproducible process to work at all. It has been a disaster because all too often the data are quite unreliable but are taken at face value anyway. Government agencies, dazzled by the high sensitivities obtainable, have written quality standards based on the technique. It is up to the instrument manufacturer to rethink the problem based on data rather than sales (26). However, the fact remains that the method is often orders of magnitude more sensitive that any other comparative technique such as plasma emission or X-ray fluorescence.

It is important to understand the flame atomizer in order to understand the value of the carbon atomizer. In the early stages of development of flame atomic absorption, it was found to be quite easy to get reliable quantitative data with a minimum of skill. The flame conditions were easily controlled, well understood, and background interference were minor. With the advent of the carbon atomizer the feeling of well-being continued and many users thought that the carbon atomizer would be just as simple to operate and get good data. In the words of Herb Kahn, then of Perkin Elmer, "The difference between flame atomizers and carbon atomizers is like the same difference as between watching tennis and playing tennis." The development of the system has been a triumph in physics because it has led to the measurement of a small signal, which is the difference between two large variable signals. In practice the variability in each signal was often greater than the signal difference between them. The difficulty was reduced by measuring the two large unreproducible signals simultaneously (or very close to simultaneously) on the same photomultiplier. If two photomultipliers were used, the difference in their responses may also be greater than the net signal being measured. A single photomultiplier must be used to measure I_0 and I_1 and "simultaneously." This has been achieved with a high degree of skill and ingenuity.

The two signals involved are the background absorption and the background plus atomic absorption signals generated when low concentrations of metals are present. Even with the three-step atomization process, it is not uncommon to reach background levels well in excess of 90%. The entire analyte signal must then be measured in the remaining absorption range, i.e. 100–90%. A 2% error in measuring either signal is a 20% error in measuring net absorption. A 2% error in measuring both background and background plus atomic absorption may be a 40% relative error. Further, the calibration curve is forced to the part of the curve where a maximum deviation from Beer's law occurs (see Ringbom plot, page 40).

In addition, the atomization process never reaches equilibrium (27). The sample is injected into the carbon atomizer and absorbed to variable extent on the surface of the carbon. The latter is then heated at a rate which is variable depending on the age and condition of the carbon atomizer. The solvent is evaporated, the remaining ash is evaporated and the remaining residue is then atomized. Unfortunately the rate of atomization depends on the original depths of penetration and rate of heating of the sample. To measure the net absorption by a population of atoms produced under nonequilibrium condition has proven to be illusive and has led to poor analytical data. It should further be pointed out that the techniques frequently used for background correction are suspect when applied to carbon atomization measurements.

Errors which can often be ignored when the background is low become unacceptable when the background is high. Using two wavelengths, a line absorbed and a line not absorbed by free atoms can only be used if the background absorption is the same for both lines (very unlikely). The use of a hydrogen lamp is acceptable if the paths of the H_2 lamp and the hollow cathode are identical. The Zeeman background corrector and the Heifje-Smith techniques measure the total signal and the background signal at times that are close together but slightly different. Considering that the total atomization time is a few seconds, this can lead to errors in calculating the areas of the signal and therefore the data.

1. Development of the Carbon Atomizer from Flame Atomizers

Analytical chemists are constantly pushing back the frontiers, and there is always a drive to obtain more sensitive procedures. The flame atomizer has been used extensively for many years and has proven capable of analyzing a great many samples submitted for routine analysis. Much research and development time has been devoted to exploiting the flame atomizer to its limit. In particular three kinds of samples were handled well by this technique: (1) the analysis of high solid-containing samples such as seawater, urine, or blood, (2) the determination of trace metals in samples available only in limited quantities such as clinical samples (3) the determination of trace metal pollutants in gases, in particular the atmosphere.

Previously these analyses included pretreatment of the sample by either selective precipitation, selective extraction, or treating gas samples with a suitable scrubbing agent.

In the case of air pollution analyses, typical concentrations of metals in air may be 1 $\mu g/m^3$ of air. If one cubic meter of air were scrubbed and the metal extracted completely, then 1 μg of that metal would be available for analysis. This in itself is a challenge, as is scrubbing a cubic meter of air, which may take several hours if it is to be done efficiently.

With this background, carbon atomization methods of analysis were developed. Mostly because of L'vov's work (28) it was recognized that they were capable of very high analytical sensitivity. At first blush it seemed relatively easy to translate this into a highly precise and reliable quantitative method of analyses, and extensive research was put into its development.

The function of the atomizer is amply described by T. S. West (29) as follows:

The basic requirements of the ideal atom reservoir are an efficient and rapid production of free atoms with the minimum background and background noise, a high level of reproducibility and a minimal memory effect and a minimal dilution of the atoms. In practice, the development of useful flame and electro thermal reservoirs is generally the result of a careful optimization compromise based on these requirements. This development process has very much favored the development and use of flame atomizers. However, recently carbon atomizers are becoming increasingly important.

2. Early Carbon Atomizers—the L'vov Furnace

The first successful electrothermal method of analysis developed for atomic absorption was described in 1961 by L'vov (28). His system is shown in the schematic diagram in Fig. 2.39. Basically the instrument was composed of two parts. The first was an electrode onto which the sample is mounted. The second was a graphite tube heated by electrical resistance. The inside of the tube was purged with argon in order to prevent oxidation of carbon surface. In addition, the

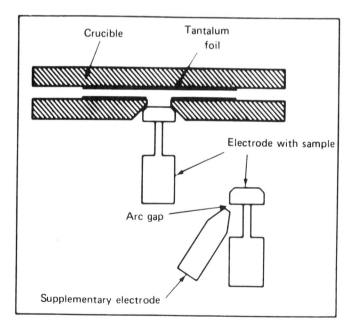

Figure 2.39 High-temperature furnace as designed and used by L'vov.

surface of the tube, both inside and outside, was treated with pyrolytic graphite to reduce diffusion through the walls. A conical hole was drilled into the side of the graphite tube, into which the sample and electrode fit. After heating up to temperature, which took between 20 and 30 sec, the sample electrode was then inserted in place and the sample vaporized. Atomization took place inside the graphite tube, and since the latter was maintained at a high temperature, the atomization was quite efficient and the atoms were maintained for a considerable time in the atomic state.

In practice, the apparatus was not easy to use, and difficulty was experienced in obtaining reproducible quantitative results. However, the sensitivities obtained by this technique were extremely good. These are listed in Table 2.19. An important effect of L'vov's work was to encourage further work in electrothermal atomic absorption research. Massman published an improved design in 1968, and at the Sheffield (England) International Conference on Atomic Absorption Spectroscopy (1969) two other papers were presented which described (1) the use of carbon filaments (29) and (2) the carbon bed atomizer (18). These papers illustrated that highly sensitive reproducible data could be achieved using carbon atomizers. Since that time several other carbon atomizers have been developed and reported upon. These will be described below.

3. The Carbon Filament Method

T. S. West and his group at Imperial College London (29) reported the development of the carbon filament method as an atomization process for atomic absorption spectroscopy. The instrument is illustrated in Fig. 2.40. Basically the atomizer was a graphite filament approximately 20 mm in length and 2 mm in diameter. The filament connected two stainless steel electrodes which were water-cooled in order to prevent them from melting in operation. A very high electrical current, about 100 A at 5 V, was passed through the filament. The latter then became very hot and thermally atomized the sample.

The first approach was to load the sample onto the filament, heat it electrically, and measure the absorption. It was very quickly realized that this technique would not work because in an extremely short period of time the solvent and other inorganic salts were vaporized, causing an extremely high background signal which completely swamped the atomic absorption signal.

The problem was overcome by West and his group, who developed a very rigorous heating program for the filament. A very small sample, e.g., 1 or 2 μl, was loaded onto the filament. The latter was then heated to a controlled low temperature for a carefully controlled period of time in order to evaporate off the solvent without sample loss. After a suitable cooling time the filament was heated again to vaporize the residue and liberate free atoms. It was found that maximum concentration of free atoms was not at the point of the highest temperature, which

Table 2.19 Sensitivity Results (1% Absorption) with L'vov Furnace

Line (nm)	Measured amount (g)	Sensitivity (g/1% absorption)
Ag 328.1	5.0×10^{-13}	1×10^{-13}
Al 309.3	2.5×10^{-11}	1×10^{-12}
Au 242.8	7.0×10^{-11}	1×10^{-12}
B 249.8	5.0×10^{-9}	2×10^{-10}
Ba 553.5	1.0×10^{-10}	6×10^{-12}
Be 234.9	2.6×10^{-12}	3×10^{-14}
Bi 306.8	2.5×10^{-11}	4×10^{-12}
Ca 422.7	2.5×10^{-11}	4×10^{-13}
Cd 228.8	6.0×10^{-14}	8×10^{-13}
Co 240.7	7.5×10^{-12}	2×10^{-12}
Cr 357.9	5.0×10^{-11}	2×10^{-12}
Cs 852.1	6.6×10^{-12}	4×10^{-12}
Cu 324.8	6.3×10^{-12}	6×10^{-13}
Fe 248.3	2.5×10^{-11}	1×10^{-12}
Ga 287.4	2.5×10^{-11}	1×10^{-12}
Hg 243.7	4.5×10^{-10}	8×10^{-11}
In 303.9	8.0×10^{-12}	4×10^{-13}
K 404.4	0.3×10^{-10}	4×10^{-11}
Li 670.8	5.0×10^{-11}	3×10^{-12}
Mg 285.2	3.0×10^{-12}	4×10^{-13}
Mn 279.5	2.5×10^{-12}	2×10^{-13}
Mo 313.5	5.0×10^{-11}	3×10^{-12}
Ni 232.0	2.5×10^{-11}	9×10^{-12}
Pb 283.3	3.0×10^{-11}	2×10^{-12}
Pd 247.6	5.0×10^{-11}	4×10^{-12}
Pt 265.9	2.5×10^{-10}	1×10^{-11}
Rb 780.0	7.5×10^{-12}	1×10^{-12}
Rh 343.5	6.3×10^{-11}	8×10^{-12}
Sb 231.1	5.0×10^{-11}	5×10^{-12}
Se 196.1	2.0×10^{-10}	9×10^{-12}
Si 251.6	2.7×10^{-12}	5×10^{-14}
Sn 286.3	1.0×10^{-11}	2×10^{-12}
Sr 460.7	2.0×10^{-11}	1×10^{-12}
Te 214.3	7.6×10^{-12}	1×10^{-12}
Ti 365.3	5.0×10^{-10}	4×10^{-11}
Tl 276.8	2.5×10^{-12}	1×10^{-12}
Zn 213.8	1.0×10^{-12}	3×10^{-14}

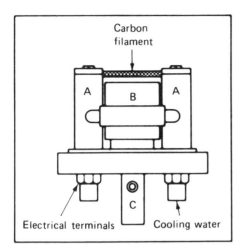

Figure 2.40 Carbon filament atom reservoir. (A) Water-cooled electrodes; (B) laminar flow box; (C) inlet for shield gas.

existed in the immediate vicinity of the carbon filament, but was some distance away. Very erratic quantitative results were obtained, and the method was highly subject to chemical interference.

The problem was reduced by sheathing the system and surrounding it with an inert gas or with hydrogen. The same atomization process was adhered to, but the inert gas or hydrogen greatly reduced the chemical interference problems and further increased the sensitivity of the procedure. An illustration of the improvement is shown in Table 2.20.

4. The Massman Furnace or Heated Graphite Atomizer

The Massman furnace was developed for commercial use by Perkin-Elmer Corporation. Basically, the atomizer is a graphite tube 50 mm long and 10 mm in diameter through which the sample beam passes. A flow of argon passes through a hollow tube, entering through five small holes and leaving through the open ends of the tube. The inert gas, usually argon, flows at a constant rate of about 1.5 liters/min. The entire tube is maintained inside a water-cooled metal cylinder.

The sample is loaded with a syringe or automatically into the center of the tube. It is then heated by a three-stage electrical heating program. These stages effect drying of the sample to remove the solvent, then ashing of the sample to remove organic material, and finally, a stronger heating step to atomize the sample.

The program must be maintained very rigorously with respect to both the temperature attained at each stage and the duration of each stage. If the ashing

Table 2.20 Interferences on Peak Height Measurement of 1 ppm of Pb

Interferent (1000 ppm)	West-type carbon rod		West-type carbon rod with H_2 flame	
	Pb, 1 ppm	No Pb	Pb, 1 ppm	No Pb
None	100	0	100	0
H_3PO_4	32	19	103	0
NaCl	110	51	92	1
$MgCl_2$	6	11	20	0
$CaCl_2$	0	46	92	1.5

Source: Ref. 15.

temperature is too high, samples can be lost due to sputtering or volatilization. The program varies from sample to sample and element to element. The manufacturers prescribe the program to be used and provide automatic devices that permit selection of the times and temperatures of each stage and automatically run the program of choice. Typical sample sizes are about 20 μl. Reported sensitivities are shown in Table 2.21. Research designs have been presented interfacing a continuous flow system with graphite furnaces (GFAAS), but they are not commercially available (30).

5. The Mini-Massman Atomizer

The carbon filament method of West was modified by Varian Associates into the Mini-Massman atomizer (Fig. 2.41). The atomizer differs from West's filament atomizer as follows: First, the geometry was changed to accommodate a small hole drilled through the rod. Into this hole the sample was placed. In a second modification, the sample was placed on top of the carbon rod. A chimney was placed beneath the rod to allow the introduction of flowing argon, nitrogen, or hydrogen which surrounded the site of atomization. The preferred sample size was of the order 0.5–1 μl.

The sample was placed into the hole using a microsyringe. The filament was subjected to a heating program that dried, ashed, and atomized. The program was automatically controlled and could be easily set on the commercial instrument.

Different programs are necessary for the quantitative determination of different elements, since the atomization rate is different for different elements, and the evaporation and ashing step also varies from one type of sample to another.

Analytical sensitivities attained by the Mini-Massman method are shown in Table 2.22.

Table 2.21   Sensitivities (1% Absorption) in the Graphite Tube Furnace (HGA-70)

Element	Absolute sensitivity (g × 10^{-12})	20 µi solution (µg/ml)
Al	150	0.007
As	160	0.008
Be	3.4	0.0002
Bi	280	0.014
Ca	3.1	0.0002
Cd	0.8	0.00004
Co	120	0.006
Cr	18	0.001
Cs	71	0.004
Cu	45	0.002
Ga	1,200	0.06
Hg	15,000	1.5
Mn	7	0.0004
Ni	330	0.016
Pb	23	0.001
Pd	250	0.013
Pt	740	0.04
Rb	41	0.002
Sb	510	0.025
Si	24	0.001
Sn	5,500	0.27
Sr	31	0.0015
Ti	280	0.014
Tl	90	0.0045
V	320	0.016
Zn	2.1	0.0001

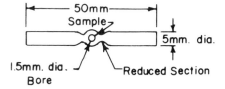

Figure 2.41   Carbon rod (Mini-Massman). Sample is located in the hole in the rod.

Table 2.22 Sensitivities (1% Absorption) and Detection Limits for Carbon Rod Atomizer (Mini-Massman)

Element	Absolute grams	Concentration (µg/ml) for 1 gml sample	Sensitivity (g/1% absorption)
Ag	1×10^{-13}	0.0002	1.2×10^{-12}
Al	3×10^{-11}	0.03	6.3×10^{-11}
As	1×10^{-10}	0.1	9.2×10^{-11}
Au	1×10^{-11}	0.01	2.1×10^{-11}
Be	9×10^{-13}	0.0009	1.1×10^{-12}
Bi	7×10^{-12}	0.007	1.0×10^{-11}
Ca	3×10^{-13}	0.0003	6.5×10^{-13}
Cd	1×10^{-13}	0.0001	6.4×10^{-13}
Co	6×10^{-12}	0.006	1.2×10^{-11}
Cr	5×10^{-12}	0.005	9.2×10^{-12}
Cs	2×10^{-11}	0.02	3.4×10^{-11}
Cu	7×10^{-12}	0.007	2.0×10^{-11}
Eu	1×10^{-10}	0.1	6.1×10^{-11}
Fe	3×10^{-12}	0.003	4.4×10^{-12}
Ga	$2 - 10^{-11}$	0.02	2.9×10^{-11}
Hg	1×10^{-10}	0.1	3.4×10^{-10}
K	9×10^{-13}	0.0009	2.0×10^{-12}
Li	5×10^{-12}	0.005	6.1×10^{-12}
Mg	6×10^{-14}	0.00006	4.3×10^{-13}
Mn	5×10^{-13}	0.0005	6.7×10^{-13}
Mo	4×10^{-11}	0.04	4.5×10^{-11}
Na	1×10^{-13}	0.0001	1.4×10^{-13}
Ni	1×10^{-11}	0.01	2.8×10^{-11}
Pb	5×10^{-12}	0.005	6.8×10^{-12}
Pd	2×10^{-10}	0.2	1.5×10^{-10}
Pt	2×10^{-10}	0.2	2.2×10^{-10}
Rb	6×10^{-12}	0.006	4.3×10^{-12}
Sb	3×10^{-11}	0.03	5.3×10^{-11}
Se	1×10^{-10}	0.1	7.2×10^{-11}
Sn	6×10^{-11}	0.06	9.6×10^{-11}
Sr	5×10^{-12}	0.005	6.1×10^{-12}
Tl	3×10^{-12}	0.003	1.1×10^{-11}
V	1×10^{-10}	0.1	9.2×10^{-11}
Zn	8×10^{-14}	0.00008	3.5×10^{-13}

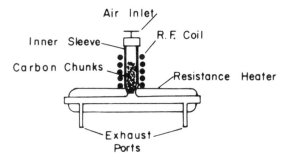

Air Inlet

Inner Sleeve

R. F. Coil

Carbon Chunks

Resistance Heater

Exhaust Ports

Figure 2.42 Radiofrequency carbon bed atomization system.

6. The Carbon Bed Atomizer Heated by Radiofrequency

Another carbon atomizer was reported at the Sheffield Conference in 1969 by Robinson (31). A feature of the system is that the background is essentially eliminated before the absorption measurement is made. A schematic diagram of the instrument is shown in Fig. 2.42. This instrument was developed originally for the direct determination of metals in air. In practice, the carbon bed was heated up to a temperature of about 1400°C with a radiofrequency coil. Air was drawn over the carbon bed and reduction took place according to the reactions:

$$C + O_2 + N_2 \rightarrow CO + N_2$$

$$MX + CO \rightarrow MO + C + \ldots$$

The technique utilizes the reducing capacities of carbon monoxide and hot carbon. In contrast to other techniques, there is no direct contact or loading of the sample onto the carbon before heating. On the contrary, the carbon bed is maintained hot. After atomization the hot gases are drawn into the horizontal T-piece. Free atoms are maintained in the light path while they are in this tube. The optical sample beam passes along this horizontal piece. Atomic absorption takes place at this stage and can be measured in the usual fashion. Solid, liquid, or gas samples could be handled directly.

7. The Hollow "T" Atomizer

The hollow "T" atomizer was a further development of the radio-frequency-heated carbon atomizer described immediately above (32). The R.F. atomizer generated a temperature of approximately 1500°C, and this was the maximum temperature possible because of the use of quartz tubes. In practice, it was found that the low temperature used was insufficient to atomize refractory elements reproducibly. The problem was overcome by using a hollow "T" atomizer made entirely of

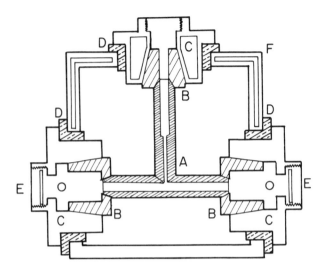

Figure 2.43   Schematic diagram of hollow "T" atomizer. (A) Graphite "T" cell; (B) graphite contacts; (C) water-cooled electrodes; (D) insulators; (E) quartz windows; (F) water-cooled housing. From Ref. 32.

carbon and heated electrically. Although it was developed independently, in many respects the design was similar to that developed by R. Woodriff in 1968 (33).

A schematic diagram of the atomizer is shown in Fig. 2.43. The atomizer was heated to a temperature of approximately 2600°C by passing an electrical current of 500 A at 12 V. An inert gas or a slightly oxidizing gas was drawn through the hollow "T" continuously at a prescribed flow rate. A liquid sample was injected into the sidearm with a microsyringe.

Combustion of the sample, including the solvent, was effected in the vertical stem. Atomization took place during this step. The products of combustion, usually CO and hydrogen, plus free atoms of the metal under consideration, wre then drawn by the flowing inert gas into the crosspiece of the "T". Here the absorption measurements were made. Sensitivities are shown in Table 2.23. They compare with other furnaces, but the accuracy and precision were much superior.

a. Molecular Interference

By burning the solvent in the side stem of the atomizer, the sample fragment and other absorbing species were broken down before reaching the light path. This eliminated the necessity for a drying and ashing step in the analytical process. Some molecular absorption from H_2 and CO was still encountered at short wavelengths (e.g., 240 nm). Using this atomizer it was possible to demonstrate that a

Table 2.23 Sensitivity of Measurement
with the Hollow "T" Atomizer

Element	Sensitivity (g/0.004 A)
As	1×10^{-10}
Be	1×10^{-10}
Cd	8×10^{-14}
Cr	8×10^{-10}
Cu	1×10^{-10}
Hg	6×10^{-10}
Mn	2×10^{-11}
Ni	4×10^{-10}
Pb	2×10^{-13}
Se	4×10^{-10}
Zn	[a]

[a]The sensitivity of zinc could not be determined
because all the samples of "pure" water obtained gave
100% absorption, and standard solutions could not be
prepared.
Source: Ref. 32.

background corrector was necessary to eliminate this source of error. Loss of volatile metals can take place during the ashing step in the Massman atomizer, as shown by Fig. 2.44.

b. Chemical Interferences

It has been shown that with the carbon filament, considerable interferences were encountered in the determination of manganese (34). The chemical interferences encountered were attributed to the varying rates of atomization of different chemical forms of the manganese. Inasmuch as the atomization process is very rapid, the rate of reaction directly affects the number of free atoms formed at any particular moment. Essentially this was a chemical interference, and directly influenced the atomic absorption measurements observed. In order to compare the hollow "T" system with the carbon filament system, a similar study was carried out on manganese. Results showed a high degree of freedom from interference which is common to determination of all metals determined with this atomizer.

c. Sensitivity

The results of a preliminary study of the sensitivities that can be achieved with the hollow "T" atomizer are shown in Table 2.23.

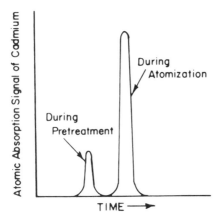

Figure 2.44   Cadmium absorption signal during a simulated dry ash atomized process.

8. Tantalum Boat Atomizers

A tantalum metal analyzer similar in most respects to the carbon filament analyzer has been developed by J. P. Matousek and B. J. Stevens (35). In this system the carbon filament is replaced by a tantalum metal strip. The heating process and heating program is the same or similar in character to that used with a carbon filament. However, the temperature is limited to 2800 K to prevent the tantalum metal from melting. The advantage of this device is that it removes any problem of carbide formation.

Each of the above devices has been used and exploited in atomic fluorescence spectroscopy. Each is capable of producing stable atom populations, and each therefore meets the prerequisite of atomizers for fluorescence work.

9. The L'vov Platform (36)

The precision of the carbon atomizer has been improved by the use of the L'vov platform, shown in Fig. 2.45. In this system a carbon platform is inserted into the standard atomizer. During the ashing and atomization step the metal atoms tend to condense on the platform, which is cooler than the electrically heated furnace.

After a short delay the platform becomes heated by radiation from the inside of the furnace, and its temperature rises. At the increased temperatures the condensed metal atoms are revaporized and entered into the light path. At this time the background has been reduced somewhat, increasing the accuracy of the method. Also, the reproducibility of the procedure is improved by the L'vov platform.

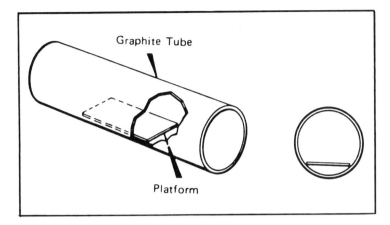

Figure 2.45 L'vov platform inside a graphite tube atomizer.

A comparison of the relative volatilization from the walls of the atomizer and from the L'vov platform is shown in Fig. 2.46. Note that the signal from the L'vov platform is greater in size; also, the delay in time helps diminish the intense background signals encountered in carbon atomizers.

Some savings can be made by using old furnace tubes. These are cut into six sections by first cutting into three "short tubes" and then slicing down the middle. One of these sections can be placed inside the furnace and acts as a L'vov furnace. It does not need to be sealed to the surface to operate (37).

10. Commercial Equipment

Commercial equipment for carbon atomization usually used Zeeman background correction or Heifje-Smith background corrector. The use of the Zeeman background corrector puts some constraints on the choice of wavelength used.

Table 2.24 shows the wavelengths used, experimental conditions, sensitivities, and other pertinent information when using the PE 5000 with a L'vov platfrom.

11. Conclusions

Thermal atomization techniques have shown themselves capable of generating very high analytical sensitivities. One of their most attractive features is the ability to handle very small sample sizes, on the order of microliters. This is particularly advantageous in clinical analyses.

For some of the techniques the problems of molecular absorption and matrix interferences are still very great, but these are much less a problem when the R.F. carbon bed atomizer is used.

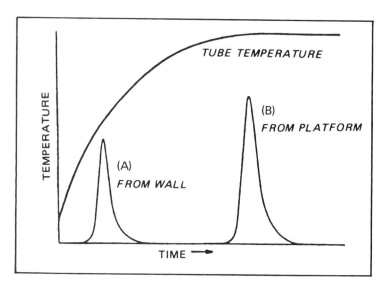

Figure 2.46 Absorption signal (A) from a sample on the wall, and (B) from a sample on the L'vov platform.

The precision of the technique is typically between 10 and 20% relative standard deviation. This may seem to be a relatively high levelof imprecision, but at concentrations of 10^{-11} and 10^{-12} g, it represents an extremely small quantity of material. Frequently, at these concentration levels analyses accurate to one significant figure are sufficient to provide the information sought.

At the present time flame atomizers still handle the bulk of samples most successfully. However, it must be conceded that other atomizers are making great strides. They have already demonstrated their capability of producing high analytical sensitivity, and witth improved atomization processes may replace the flame as the principal atomization process used in routine analysis.

One problem with the carbon atomizer is the formation of carbides by some elements, particularly aluminum, silicon, tungsten, and boron (38). The problem has been somewhat overcome in the case of aluminum and silicon, but not for tungsten or boron.

A problem with this technique was that the sample soaked into the carbon rod, facilitating the formation of carbides. The use of pyrolytic graphite to coat the surface of the carbon rod greatly reduces this problem by reducing the porosity of the graphite surface.

A survey of recent research in the fields of atomicc spectroscopy has been carried out by Holcombe and Bass (39). In their discussion on atomic absorption, atomic fluorescence, and flame emission, they quoted a total of 1137 references. Clearly the field is still very active and vibrant.

Table 2.24 Zeeman Background Correction Data for PE 5000, 1988 (Courtesy of Perkin-Elmer Corp.)

Element	λ (nm)	Slit (nm)	Site[a]	Modifier	Pretreat. (°C)	Atom. (°C)	Detection Limits (pg)	Detection Limits (μg/L)[c]
Ag	328.1	0.7	P	0.01 mg Pd[b]	1100	1600	0.5	0.005
Al	396.2	0.7	P	0.005 mg Mg(NO₃)₂	1700	2500	5.0	0.05
As	193.7	0.7	P	0.02 mg Ni	1300	2300	20.	0.2
Au	242.8	0.7	P	0.05 mg Ni	1000	2200	10.	0.1
B	249.7	0.7	W	0.005 mg Ca	1000	2650	1000.	10.
Ba	553.6	0.2	W		1200	2500	10.	0.1
Be	234.9	0.7	P	0.05 mg Mg(NO₃)₂	1500	2500	0.5	0.005
Bi	223.1	0.2	P	0.02 mg Ni	900	1900	10.	0.1
Ca	422.7	0.7	W		1100	2600	1.	0.01
Cd	228.8	0.7	P	0.2 mg PO + 0.01 mg Mg(NO₃)₂	900	1600	0.2	0.002
Co	242.5	0.2	P	0.05 mg Mg(NO₃)₂	1400	2500	5.	0.05
Cr	357.9	0.7	P	0.05 mg Mg(NO₃)₂	1650	2200	2.	0.02
Cs	852.1	0.7	P	0.2% H₂SO₄	900	1900	3.	0.03
Cu	3724.8	0.7	P		1000	2300	2.	0.02
Dy	421.2	0.2	W		1500	2650		
Er	400.8	0.2	W		1700	2650		
Eu	459.4	0.2	W		1300	2600		
Fe	248.3	0.2	P	0.05 mg Mg(NO₃)₂	1400	2400	2.	0.02
Ga	294.4	0.7	P	0.05 mg Mg(NO₃)₂	1200	2000	7.	0.07
Gd	407.9	0.2	W		1600	2650		

(Continued)

Table 2.24 (*Continued*)

Element	λ (nm)	Slit (nm)	Site[a]	Modifier	Pretreat. (°C)	Atom. (°C)	Detection Limits (pg)	Detection Limits (μg/L)[c]
Ge	265.1	0.2	P	0.05 mg Mg(NO$_3$)$_2$	950	2400	15.	0.15
Hg	253.7	0.7	P	0.02 mg Pd[b]	140	2000	40.	0.4
In	325.6	0.7	P	0.015 mg Pd + 0.01 mg Mg(NO$_3$)$_2$	1200	2100	5.	0.05
Ir	264.0	0.2	W		1300	2650		
K	766.5	0.7	W		950	1500	0.5	0.005
La	550.0	0.2	W		1600	2650		
Li	670.8	0.2	P		900	2600	2.	0.02
Mg	285.2	0.7	P		900	1700	0.4	0.004
Mn	279.5	0.2	P	0.05 mg Mg(NO$_3$)$_2$	1400	2200	1.	0.01
Mo	313.3	0.7	W		1800	2650	4.	0.04
Na	589.0	0.7	W		900	1500	1.	0.01
Nd	463.4	0.2	W		1500	2650		
Ni	232.0	0.2	P		1400	2500	10.	0.1
Os	290.9	0.2	W		200	2650		
P	213.6	0.7	P	0.02 mg Pd + 0.005 mg Ca	1400	2650		
Pb	283.3	0.7	P	0.2 mg PO$_4$ + 0.01 mg Mg(NO$_3$)$_2$	850	1800	5.	0.05
Pd[b]	247.6	0.7	P		900	2650	25.	0.25
Pt	265.9	0.7	W		1300	2650	50.	0.5

Rb	780.0	0.7	P		800	1900	2.	0.02
Rh	343.5	0.2	W		1300	2400	10.	0.1
Ru	349.9	0.2	W		1400	2500		0.2
Sb	217.6	0.7	P	0.02 mg Ni	1100	2400	20.	0.2
Se	196.0	2.0	P	0.02 mg Ni + 0.025 mg Mg(NO$_3$)$_2$	900	2100	20.	0.2
Si	251.6	0.2	P		1400	2650	30.	0.3
Sm	429.7	0.2	W		1400	2600		
Sn	286.3	0.7	P	0.2 mg PO$_4$ + 0.01 mg MG(NO$_3$)$_2$	800	2100	20.	0.2
Sr	460.7	0.7	W		1300	2600	2.	0.02
Te	214.3	0.2	P	0.02 mg Ni	1000	2000	10.	0.1
Ti	stet	0.2	W		1400	2650	40.	0.4
Tl	276.8	0.7	P	0.015 mg Pd + 0.01 mg Mg(NO$_3$)$_2$	900	1600		
Tm	371.8	0.2	W		1700	2650	5.	0.05
U	351.5	0.2	W		1000	2650		
V	318.4	0.7	W	0.05 mg Ng(NO$_3$)$_2$	1100	2650	20.	0.2
Y	410.2	0.2	W		1400	2650		
Yb	398.8	0.2	W		1300	2500		
Zn	213.9	0.7	P	0.006 mg Ng(NO$_3$)$_2$	700	1800	1.	0.01

[a] P = platform; W = wall atomization.
[b] Pd is pretreated at 1000°C before adding sample.
[c] Assuming 100-μL sample aliquot.

Another factor is the chemical form of the sample in the solvent. If the solvent is aqueous, then usually the sample is ionic. On the other hand, if the solvent is organic, then generally the metal exists as a metal-organic compound. When the solvent is evaporated off completely, we are left with a residue containing the metal of interest. If it is a metal ion salt, then energy is again required to break it down to the free neutral state. However, if it is a metal-organic compound, then the organic addend is combustible and will react in the flame and release the free neutral atoms more readily than in the case of the inorganic metal salt.

The process of combustion of the solvent in the flame is illustrated in Table 2.3. It we insert illustrative numerical values for the number of atoms for each step in the process, we can see that organic solvents cause an increase in the number of excited and unexcited atoms produced by this system.

REFERENCES

1. Walsh, A., *Spectrochim Acta 7*: 108 (1955).
2. Hamly, J., Multielement Atomic Absorption with a Continuous Source, *Anal. Chem.58, No. 8*: 933A (1986).
3. Jones, W. G., and A. Walsh, *Spectrochim Acta, 16*: 249 (1960).
4. Robinson, J. W., and E. M. Skelly, *Spec. Letters, 14 (7)*: 519 (1981).
5. Robinson, J. W., *Indus. Chem.*, 5: 225 (1962).
6. *Improved Hollow Cathode Lamps for AAS*, S. Caroli, ed. John Wiley and Sons, New York, 1985.
7. Robinson, J. W., and R. J. Harris, *Anal. Chim. Acta, 26*: 439 (1962).
8. Backstone, K., and L. Davidson, *Anal. Chem, 60*: 1354 (1988).
9. Martin, J. M., and T. J. Ihrig, *App. Spec., 41 (6)*: 986 (1987).
10. Copeland, T. R., K. W. Olson, and R. K. Skoyerboe, *Anal. Chem.*, 44: 1471 (1972).
11. Robinson, J. W., and J. C. Wu, *Spec. Letters, 18 (6)*: 399 (1985).
12. Blakely, C. R., and M. L. Vestal, *Ann. Chem.*, 55: 750 (1983).
13. Robinson, J. W., and D. S. Choi, *Spec. Letters, 20 (4)*: 375 (1987).
14. Rain, C. S., and A. N. Hambly, *Anal. Chem.*, 37: 879 (1985).
15. Amos, M. D., *Spectrochim Acta, 22*: 1325 (1966).
16. Willis, J. B., *Nature, 207*: 715 (1965).
17. Gaydon, A. G., and H. G. Wolfhand, *Flames*, Chapman and Hall, 1960.
18. Hadeishi, T., and R. D. McLaughlin, *Science, 174*: 404 (1971).
19. Brech, F., *L.S.U. Symposium in Analytical Chemistry*, 1974.
20. *Atomic Spectra*, C. Chandler, Van Nostrand, 1964, p. 64.
21. Brodie, K. G., and P. R. Liddell, *Anal. Chem.*, 1980, *52*, 1059.
22. Smith, F. B., Jr. and G. M. Hiefje, *Applied Spectroscopy, 37 (5)*: 419 (1983).
23. Camrick, G. R., and W. Slavin, *Am. Lab.*, (Feb.): 90 (1989).
24. Manny, D. C., and W. Slavin, *Applied Spec., 37 (1)*: 1 (1983).
25. McGregor, D. A., K. B. Cull, J. N. Gehlhausen, A. S. Viscomi, M. Wu, L. Zhang and J. W. Carnahan, *Anal. Chem.*, 60 (19): 1089A (1988).
26. Vellon, C., *Anal. Chem.*, 58 (8): 851A (1986).

27. Robinson, J. W., and P. J. Slevin, *Amer. Lab.*, (Aug.): 10 (1971).
28. L'vov, B. V., *Spectrochim Acta, 17*: 761 (1961).
29. West, T. S., Sheffield International Conference on Atomic Spectroscopy, 1969.
30. Backstrom, K., and L. Danielson, *Anal. Chem. 60*: 1354 (1988).
31. Robinson, J. W., Sheffield International Conference on Atomic Spectroscopy, 1969.
32. Robinson, J. W., W. Wolcott, *Anal. Chim. Acta, 74*: 43 (1975).
33. Woodruff, R., R. W. Stone and A. M. Held, *Appl. Spec.*, (1986).
34. Kirkbright, G. F., et. al., *Anal. Chim. Acta, 39*: 58 (1972).
35. Matousek, J. P., and B. J. Stevens, *Anal. Chem., 17*: 363 (1971).
36. Manny, D. C., and W. Steenan, *Applied Spec., 37 (1)*: 1 (1983).
37. Rains, T., private communication, 1988.
38. Styris, D. L., and R. A. Redfield, *Anal. Chem., 59*: 2891 (1987).
39. Holcombe, J. A., and D. A. Bass, *Anal. Chem., 60*: 226R (1988).

3

Atomic Fluorescence

A. INTRODUCTION

The basic process of atomic fluorescence is (A) the creation of free atoms from the sample, (B) the absorption of radiation by these free atoms, which become excited in the process, and (C) the reemission of radiation when the atoms return from the excited state to a lower energy state—usually the ground state. Two steps of this process are similar in many aspects to atomic absorption spectroscopy, that is, the creation of free atoms from a sample and the process of absorption of radiant energy by these atoms. The third step is similar in many respects to flame emission spectrophotometry.

Atomic fluorescence from excited sodium was first observed in 1905 by Wood (1) and has been observed by numerous other workers since then. However, it was essentially a field of "academic interest only" until it was used by Robinson to demonstrate that atomic excitation by radiation directed at a flame was possible (2) and that thermal excitation as defined by the Boltzmann distribution was only part of the atomic excitation process. At that time atomic fluorescence was not pursued as a quantitative method of analysis since the process would probably suffer from the interference of both atomic absorption and flame photometry.

Winefordner eliminated some of these interferences by modulating the equipment. Rigorous studies were initiated and reported, first by Winefordner, Vickers, and Staab (3, 4), and then by a group spearheaded by T. S. West (5). Both research groups proposed that atomic fluorescence held some advantages over atomic absorption spectroscopy, particularly the fact that the fluorescence signal was proportional to the intensity of radiation that excited the sample. They showed that using an extremely strong excitation radiation, such as a tunable laser, the fluorescence signal similarly increased and that there was virtually no theoretical limit to the intensity of the fluorescence signal that may thus be produced. This projection was in direct contrast to atomic absorption spectroscopy, where the absorption signal being measured is the fraction of the radiation falling on the atom population. The absorption signal (% absorption) is essentially independent of the radiation power, hence it is not increased by increasing the power of the radiation.

In addition, in atomic absorption spectroscopy it is vital that the line width of the radiation source be at least as narrow and preferably narrower than the atomic absorption line width in order to effect maximum absorption of the line source of radiation. In contrast, in atomic fluorescence the radiation light source does not fall on the detector, and therefore light sources of any available line width or intensity can be used. It is an advantage if the light source is broad relative to the absorption band line width, since this ensures the maximum amount of absorption of radiation by the atom population. This results in an increased population of excited atoms and, therefore, an increased fluorescence signal.

B. MATHEMATICAL RELATIONSHIPS

The mathematical relationships governing fluorescence intensities have been adequately covered by Winefordner and Vickers (4). The following is based on their treatment.

From Beer's law, $T = \frac{I_1}{I_0}$, $A = -\ln T = abc$,

$$T = e^{-abc}$$

$$\frac{I_1}{I_0} = e^{-abc}$$

$$1 - \frac{I_1}{I_0} = 1 - e^{-abc}$$

Multiply both sides by I_0:

$$I_0 - I_1 = I_0(1 - e^{-abc}) \tag{3.1}$$

where $I_0 - I_0$ = amount of light absorbed.

The intensity of fluorescence P_F is proportional to the quantity of radiant energy absorbed. Therefore,

$$P_F = (I_0 - I_1)\phi$$

$$P_F = P_{abs}\phi$$

where

P_F = intensity of fluorescence (total)

P_{abs} = quantity of radiant energy absorbed

ϕ = quantum efficiency of the process, that is, the number of atoms that undergo observed fluorescence transformation per unit time divided by the number of atoms excited from state 1 per unit time.

From Eq. (3.1) the relationship between the absorbed and the incident radiation is given by the expression:

$$P_F = I_0(1 - e^{-abc})\phi \tag{3.2}$$

If abc is small,

$$P_F = KI_0 \, 2.3 \times abc\phi \tag{3.3}$$

where K is an instrumental factor

Therefore,

$$P_F = KI_0 c$$

where K includes k, a, b, and 2.3—all of which are constants.

For atoms, this can be written as

$$P_F = P_0(1 - e^{-k_0 Lc})\phi$$

where

P_0 = intensity of incident radiation

k_0 = atomic absorption coefficient

L = length of the absorption cell

Δv = half the base width of the absorption profile

If an absorption line is scanned from a point remote from its center, absorption is zero. It increases to a maximum as the center of the line is reached. As the scan leaves the center of the line, the absorption again diminishes to zero. The shape of the curve relating absorption and wavelength is the absorption profile Δv of the line.

An approximation to Δv can be made using the approach made by Willis (6). If the absorption profile is gaussian in shape, it can be approximated to a triangle. Under these circumstances, we have

$$\Delta v = \frac{\pi}{\ln^2} \tag{3.6}$$

where ΔvG is the absorption line spectral width at a half intensity of the gaussian curve. This quantity can be measured. By combining Eqs. (3.2) and (3.4) we have

$$P_F = P_{abs}\phi$$

$$= \phi \, P_0 \Delta v \, (1 - e^{-koL}) e^{-koL/2} \cosh \left(\frac{K_0 L}{2} \right)$$

$$= CP_0 N$$

N = Number of atoms in the light path

where $\cosh (K_0 L/2)$ is a correction term to accommodate self-absorption by the sample (8).

Equation (3.7) can be rewritten in terms of the intensity measured, I_F, by dividing P_F by the viewed area A_F of the cell (flame) and by 4π steradians. Then

$$I_F = \frac{\phi P^0 \Delta v k^0 \, L}{4\pi A_F} \tag{3.8}$$

But earlier work by Mitchell and Zemansky indicated that

$$k^0 = \frac{(\ln 2)^{1/2} \, \lambda^2 g_1}{4\pi^{3/2} \, \Delta v_D g_2} \, NA \, \delta \, (av) \tag{3.9}$$

where

Δ_D = Doppler half-width

g_1/g_2 = a priori statistical weights of atoms in states 1 and 2

A_t = transition probability

λ = wavelength of the absorbing line

$\delta(av)$ = rate change of damping coefficient cm^{-1}

By substituting Eqs. (3.9) and (3.8), we have

$$I_F = \frac{(\ln 2)^{1/2} \phi \Delta v \, L \lambda^2 A_f \, \delta(av) g_1}{16\pi^{5/2} A_F \, \Delta v_D g_2} \tag{3.10}$$

$$= CP_0 \, N$$

where C is a constant for any particular experimental arrangement. This leads to the relationship that the intensity of fluorescence is proportional to the incident radiation and the number of atoms N that can absorb.

This relationship ignores self-absorption at higher concentrations (see Fig. 3.1). A correction for self-absorption leads to the relationship:

$$I_F = kc - kc^2$$

and a higher concentration reversal is apparent (Fig. 3.1).

Many of the factors that affect atomic absorption will affect atomic fluorescence. This includes N, the number of atoms in the light path, and f, the oscillator strength of the absorption line. All the factors that affect N in atomic absorption should affect N in atomic fluorescence in a similar fashion, and should include chemical interferences, solvent effects, atomization efficiency, flame composition, and the stability of the neutral atoms.

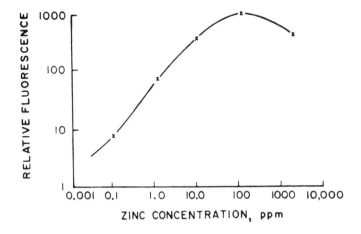

Figure 3.1 Relative fluorescence of Zn (213.9 nm) at different concentrations (13). Note reversal of Zn fluorescence in concentrations greater than 100 ppm.

In contrast to atomic absorption, the degree of fluorescence of excited atoms should depend on the quantum efficiency of the process. This in turn is dependent on the efficiency of fluorescence, deactivation by other means (for example, collision), or the loss of radiant energy by other processes in the atom.

C. ADVANTAGES OF ATOMIC FLUORESCENCE

The attractive features of atomic fluorescence include the following:

1. I_F, the fluorescense intensity, can be increased by increasing P_0, the incident radiation.
2. C [Eq. (3.10)] can be increased by increasing L, the size of the flame, and the quantum efficiency.
3. N, the number of fluorescing atoms, is a function of the unexcited neutral atoms in the system. This is inherently higher than the number of thermally excited atoms.
4. A radiation source with wide spectral lines can be used.
5. The intensity of fluorescence is linearly related to the concentration of the sample element over a wide concentration range.
6. The element may be excited at one wavelength and the fluorescence measured at a different wavelength. This eliminates the effect of scattered radiation.
7. Theoretical sensitivity limits are claimed by some to be higher than other conventional atomic spectroscopic methods.

D. LIMITATIONS OF ATOMIC FLUORESCENCE

A serious disadvantage of atomic fluorescence is self-absorption of the fluorescence by the sample. This leads to a reversal of the slope of the curve relating I_F, the fluorescence intensity, and the analysis of the sample at high concentrations. It can be overcome by suitable dilution of the sample.

The limitations of atomic absorption also apply. If a flame atomizer is used, metals that form refractory oxides will be difficult to detect; also, with the present development of equipment, only one element can be determined at a time. The fluorescence intensity may suffer from background interferences by the atomizer (flame).

The intensity of fluorescence of a particular line can be affected by four types of fluorescence. Of these, sensitized fluorescence in particular can cause a direct interference to the intensity.

E. ATOMIC FLUORESCENCE AS AN ANALYTICAL TOOL

1. Atomization Process

As in atomic absorption spectroscopy, a major step in atomic fluorescence is the generation of an atom population from the original sample. As in atomic absorption spectroscopy, the most common atomizer used has been the flame.

These systems have been described earlier. Other atomization systems have been developed and successfully demonstrated for atomic fluorescence. These include separated flames, the Massman furnace, and the carbon filament. The latter two were described in Chapter 2.

a. Separated Flame Atomizers.

It will be remembered from Chapter 2 that the flame consists of three major sections: the reaction zone, the inner core, and the outer core. In the reaction zone the major part of combustion and atom reduction takes place. In the inner core further reduction and combustion takes place, and here many free atoms exist. In the outer core air is entrained and combustion is complete, but frequently, free atoms are lost.

It is a distinct advantage to be able to extend the inner core in order to produce a stable population of atoms not disturbed by entrainment of the air. The separated flame, developed by T. S. West, was particularly valuable when used to this end. Earlier designs used quartz tubing placed around the flame, which separated the reaction zone from the outer mantle. A later design used an inert gas to separate the flame from the ambient air. A schematic diagram of this burner is shown in Fig. 3.2. An advantage of this system is that the intense radiation from the reaction zone is removed from the optical system. This permits the use of solar-blind detectors and greatly increases the detection limits of atomic fluorescence.

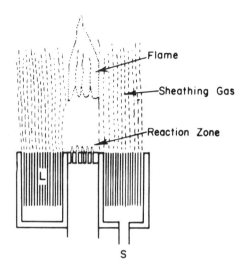

Figure 3.2 Separated flame for atomic fluorescence. L = capillaries for laminar gas flow: S = sheathing gas inlet. After West and Cresser (6).

The most common flame is the air–acetylene flame, although the nitrous oxide–acetylene flame has been used extensively, as have oxyhydrogen flames. Argon is the commonly used as the sheathing gas.

It will be remembered that the atomization efficiency depends on flame temperature, flame composition, the solvent, the principal anions (and sometimes cations), and the position in the flame. The discussions in Chapter 2, Sec. E should be reviewed at this point.

Atomic fluorescence has also been used in conjunction with a tube electrothermal atomizer (ETA) to produce high sensitivity. The fluorescence was generated by laser excitation (LEAFS). A Zeeman background corrector was used, and several commercial atomizers were examined. The sensitivities were high, as shown in Table 3.1.

2. The Absorption Process

Absorption of radiant energy is the process by which the atoms become excited. For exactly the same reason as those considered in atomic absorption, the important energy treansitions are transitions from the highly populated ground state to higher excited states. Transitions between upper excited states are possible, but the very low population of these states results in very low analytical sensitivity.

Table 3.1   Comparison of Detection Limits (pg)

Element	This work ETA LEAFS (10)		lit. ETA LEAFS	lit. ETA AAS
	On line	Off line		
Ag	0.02		0.1	0.5
Co	0.3		0.06	2
Cu	0.6		0.2	1
In	0.02		0.1	9
Mn	0.4	0.1	0.2	1
Pb	0.2	0.01	0.002	5
Tl	0.1		0.003	10

The quantity of light absorbed depends on the number of absorbing atoms in the light path and the efficiency of these atoms in absorbing radiation, i.e., the oscillator strength. The oscillator strength is a physical property of the particular atomic species and the energy levels concerned (see page 90). The number of atoms in the light path depends on the atomization process and is subject to the same interferences as those encountered in atomic absorption spectroscopy, particularly *chemical interferences*.

It should be pointed out that the quantity of light absorbed $(I_0 - I_1)$ is increased and therefore a greater number of atoms will be excited when the intensity of radiation falling on the population is increased, even though the fraction of radiation absorbed $\dfrac{I_0 - I_1}{I_0}$ remains the same. It is therefore very much an advantage to use highly intense light sources such as laser beams. This is in contrast to atomic absorption spectroscopy, where the fraction of light absorbed (I/I) rather than the total amount of light absorbed $(I-1)$ is the important factor.

In the excitation process, all the empty orbitals at higher energy levels are available to be filled by an excited electron. In practice, only the first excited state and the second or third excited state are of analytical value in atomic fluorescence. This is because the oscillator strength of transitions between the ground state and higher excited states is comparatively low, and therefore the quantity of radiation absorbed greatly decreases, resulting in a weak fluorescence signal.

3. The Fluorescence Process

In molecular fluorescence the absorbing and fluorescing species is a molecule. Like molecules, atoms are also capable of being electronically excited and subsequently fluorescing (2). However, molecules include complicated geometric forms that exhibit vibrational and rotational energy levels. Excited molecules may be unstable and lose their electronic energy to other energy forms such as heat or vibrational or rotational energy. These cause the molecular fluorescense to be broad band. In contrast, the atomic fluorescence lines are very narrow, since they are not broadened by vibrational or rotational bands.

4. Types of Fluorescence

Winefordner and Vickers (3) suggest that there are four principal types of fluorescence These they defined as follows.

a. Resonance Fluorescence

Resonance fluorescence occurs when the absorption wavelength and the fluorescence wavelength are identical. The transitions are illustrated in Fig. 3.3. For all practical purposes it refers to the use of atomic lines associated with the transition between the ground state and the first excited state of the valence electron.

Three difficulties are involved with resonance fluorescence. First, any scattered radiation of the light source by particles in the flame is at the resonance fluorescence wavelength. Scattered radiation cannot be distinguished from resonance fluorescence since both are at the same wavelength. It is not eliminated by modulation. This results in a direct analytical interference giving falsely high results.

A second problem with resonance fluorescence is that frequently the atom may be excited to the higher (e.g., second) excited state, lose energy, and drop down to the first excited state as in stepwise fluorescence (see Sec. D.3.a). Having reached the first excited state, it fluoresces and descends to the ground state. The wavelength of this fluorescence is identical to resonance fluorescence and results in an enhancement of the resonance fluorescence intensity. It is a direct analytical interference unless careful steps are taken in the calibration procedure to accommodate this source of radiation.

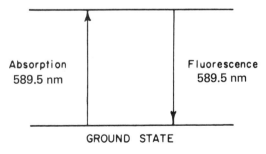

Absorption
589.5 nm

Fluorescence
589.5 nm

GROUND STATE

Figure 3.3 Resonance fluorescence of sodium at 589.5 nm.

A third problem is self-absorption of the fluorescence signal. An atom deep within the flame may fluoresce at the resonance wavelength, but an emitted photon may be reabsorbed by nearby unexcited atoms before it leaves the flame. It does not reach the detector, and the measured fluorescence intensity is decreased. The problem becomes more severe at higher sample concentrations where measured fluorescence intensity may actually decrease with increase in concentration (line reversal). The use of calibration curves helps to offset this problem but must be used with care. Successive dilution of the sample can be used to indicate if the intensity reading is on the linear part of the curve or the reversed region at high concentration.

The principal advantage of resonance fluorescence is that it gives the *most intense fluorescence* signal and provides the highest analytical sensitivity. However, it is also the line most subject to analytical interferences and therefore is most difficult to deal with in attempting to get reliable quantitative data.

b. Direct-Line Fluorescence

Direct-line fluorescence is a process in which the valence electron is excited from the ground state to a higher excited state. From this high excited state it falls to a lower excited state; in the process it fluoresces. The fluorescence is at a longer wavelength then the absorption wavelength (Fig. 3.4). The advantage of using direct-line fluorescence is that it eliminates interferences encountered with resonance fluorescence, such as scattered radiation and increased excited atom population.

However, it should be pointed out that the oscillator strength of the absorption transition between the ground state and the second excited state is lower than that between the ground state and the first excited state. This results in fewer excited

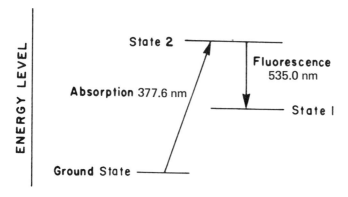

Figure 3.4 Direct-line fluorescence of thallium.

atoms being available for fluorescence. In addition, only a small fraction of these excited atoms will descend to the first excited state prior to fluorescence rather than to the ground state; hence, the process is quite inefficient.

Direct-line fluorescence is of reduced analytical sensitivity but is better for quantitative determinations. This is particularly so when the sample concentration is high enough to provide a reasonably intense fluorescence signal. It should also be pointed out that the transition involved—that is, between the second and first excited state— is one that is also common to the metal in the source, and it is very likely that this line will be an emitted line from the source itself. Unless steps are taken to prevent this radiation from reaching the atomizer, it can be scattered in the same way as the resonance line, resulting in a falsely high fluorescence signal. The problem can be overcome by inserting a filter between the light source and the atomizer which does not permit radiation of this wavelength from the source to reach the atomizing system. This in turn prevents scattering of the radiation and eliminates scatter as a source of error.

c. Stepwise Fluorescence

In stepwise fluorescence the valence electron is excited to a higher energy level in the same was as in a direct-line fluorescence. Here the excited electron loses energy in a nonradiative manner and descends to a lower excited state. Having reached the lower excited state, it then fluoresces and returns to the ground state (Fig. 3.5).

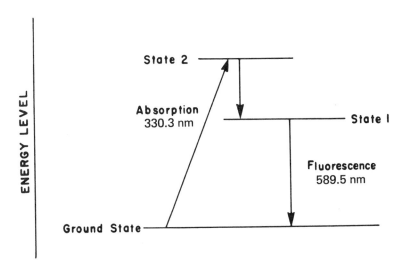

Figure 3.5 Stepwise fluorescence of sodium.

Analytically, this fluorescence is of distinctly lower intensity than resonance fluorescence, even though it occurs at the same wavelength as resonance fluorescence. This is because the oscillator strength of the transition between the ground state and the higher excited state is lower than in resonance line absorption, resulting in a decreased number of excited atoms. Of these excited atoms, only a fraction will descend to a lower excited state by a nonradiative process. This results in a relatively low population of atoms in the excited state from which fluorescence occurs. As a result, the analytical sensitivity is significantly lower than that observed in resonance fluorescence. In addition, this line is at the same wavelength as resonance fluorescence, and therefore steps must be taken to prevent scattering of the resonance line which will be strongly emitted if a hollow cathode light source is used. As in direct-line fluorescence, this source of interference can be eliminated by inserting a filter between the hollow cathode and the atomizer.

A second analytical interference is an increased population of the pertinent excited state from atoms in an even higher excited state. This results in an increase in the emission intensity and an apparent increase in analytical sensitivity. Unless care is taken in the calibration procedure, a direct analytical interference will occur. Modulation removes this interference with all lines.

d. Thermally Assisted Fluorescence

In thermally assisted fluorescence the excitation procedure is little more complicated than in the previous examples. First, the atom is excited by absorption of radiation from the ground state to an upper excited state. Here the already excited atom is further excited by thermal collision to an even higher excited state. This process was first observed in 1967 by West and his group (9). The lines are quite weak but are of academic interest.

e. Sensitized Fluorescence

In sensitized fluorescence, excitation first takes place by collision activation to an excited state followed by fluorescence back to the ground state. An example of this process is the behaviour of a mixture of mercury plus thallium vapor. When irradiated at the mercury resonance line at 253.7 nm, the mercury atoms become excited. These excited atoms collide with thallium atoms, which then become excited in the process. The latter fluoresce at Tl 377.6 nm and 535 nm. This is an unusual process and not of practical importance in atomic fluorescence. It is not usually observed using flame atomizers. It is of course a potential source of interference and should be remembered when setting up quantitative analytical procedures.

5. Equipment

All the early work in atomic fluorescence used equipment based on the optical systems and components of flame photometry or atomic absorption instrumenta-

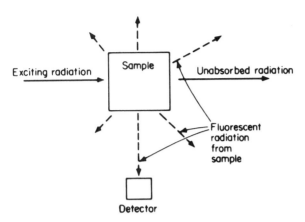

Figure 3.6 Schematic diagram of fluorescence equipment.

tion. These are illustrated in a schematic diagram of instrumentation (Fig. 3.6). High-intensity hollow cathode lamps were used as the light source, because they provided the maximum light power available for exciting the atoms. The components of the equipment were similar to those used in flame photometry or atomic absorption. In principle, the sample was atomized using a flame atomizer, the free atoms were excited using a hollow cathode lamp, and the fluorescence was measured at right angles to the excitation source beam. The monochromator was used to select the pertinent wavelength for measurement.

Most of the components are the same as those used in atomic absorption spectroscopy and will not be discussed further in this section. In particular we shall not discuss the atomizer, the monochromator, and the detector readout system since these are discussed in Chapter 2.

Early equipment used line-source radiation sources such as hollow cathodes, but better sensitivity can be obtained using a tunable laser. Continuous radiation sources such as hydrogen lamps have been suggested because all wavelengths are excited simultaneously. One of the serious disadvantages of continuous radiation sources is the problem of ligh scattering. Since the source emits over a wide wavelength range, any scattering that takes place in the atomizer (such as the flame) will also take place over a wide spectral range. Unless steps are taken to eliminate or correct for this signal, a serious analytical error will result.

Early work on atomic fluorescence was devoted to the use of continuous sources. However, there has been a decrease in attention to this type of source because of the difficulties involved in getting quantitative analytical data. But its initial potential still remains, in that it is a simple source and has the potential of exciting many elements in the Periodic Table. An evaluation of continuous and line excitation sources has been made by Perkins and Long (9).

a. Radiation Sources

Two principal types of sources used for atomic fluorescence are continuous radiation sources and line radiation sources. In the former, radiation occurs over a wide band, and such sources are potentially useful for excitation of all the metals in the Periodic Table. Radiation line sources are suitable for the excitation of the particular element used to generate the line.

The basic premise of the use of line sources is that they emit radiation at precisely the wavelength that causes excitation of the element under consideration. Considerable early developmental work was devoted to the use of metal vapor discharge lamps succh as mercury lamps, hollow cathode lamps, high-intensity lamps and electrodes discharge lamps.

It was found that increased light intensity resulted in an increase in sensitivity, but not necessarily improved in acccuracy or precision. A common approach is to use a pulsed light source such as a pulsed hollow cathode lamp or a pulsed laser. Pulsing can be done by using mechanical chopper or an a/c power supply. The pulsing of the light source causes a corresponding pulsing of the fluorescence signal. This is exactly analogous to modulating the signal. Many interferences are eliminated, particularly flame emission, both background and line emission (as in flame photometry) arising at the same wavelength as the fluorescence. This is particularly important when the resonance lines are monitored.

Tunable dye lasers are used such as the laser using rhodamine 6G, which is widely used for studies on sodium fluorescence. Using very strong sources, such as YAG-pumped dye laser, the fluorescence signal becomes saturated and no longer proportional to the light intensity I_0. This ensures maximum sensitivity and makes the signal insensitive to small variations in laser intensity. Some systems have been combined with an AA carbon atomizer for improved sensitivity (10).

b. The Electrodeless Discharge Lamp (EDL)

The electrodeless discharge lamp (EDL) was considered to be a radiation source well worth study. This lamp is described below. The lamp blank for an electrodeless discharge lamp is successively flushed with argon and evacuated, then charged with suitable metal, recharged with argon at atmospheric pressure or higher, and sealed off. The lamp is then inserted inside a microwave discharge cavity. Here it is excited by the microwave radiation and emits at the emission spectrum including the resonance line of the element inside the sealed lamp blank.

The intensity of these lines is considerably greater than that of the high-intensity hollow cathode lamp. One major problem with this source is its instability. It is vital to quantitative analysis that the lamps be very stable in order to correlate fluorescence intensity with the concentration of the element being determined.

c. Modulation of Equipment

In order to avoid error caused by background emission from the flame as opposed to fluorescence, it is necessary to modulate the equipment. This can be done with a mechanical chopper or by electronic means. It should also be pointed out that modulation is vital in order to avoid the interferences that would result from atomic radiation from the sample element present in the flame. The thermal emission of the sample atoms in the flame occurs at exactly the same wavelength as those used in atomic fluorescence, and unless they are corrected for they will result in a direct analytical interference to quantitative analysis. The process of modulation has previously been described (see Chapter 2).

d. Nondispersive Atomic Fluorescence Systems

One of the virtues of atomic fluorescence is that the fluorescence to be monitored can be excited selectively. This is in contrast to flame photometry, where a broad, rich spectrum is emitted by the flame, and to atomic absorption, where a similar rich, intense spectrum is emitted by the hollow cathode discharge lamp. In each of the latter techniques the spectrum falls upon the detector unless special care is taken to select the wavelength of interest using the monochromator.

In contrast, in atomic fluorescence the wavelength of interest can be excited by a suitable radiation source. Fluorescence at the fluorescing and other wavelengths occur and can be measured at right angles to the excitation beam. If the wavelength of this fluorescence is less than 350 nm, then the background emission of the flame is very low and a nondispersive optical system can be used.

Such a system was greatly improved by the advent of solar-bind detectors. the latter are insensitive to visible radiation, and since it is at these wavelengths that most flames are very intense, the very strong background signal from the flame is not registered by the detector. The detector, therefore, registers only radiation at short wavelengths.

A good illustration is the fluorescence of zinc at 213.8 nm. Here the detector can monitor only radiation between 200.0 nm and 300.0 nm. The zinc may be atomized and excited and will fluoresce at 213.8 nm. The detector registers only the increased signal falling upon it, which in this case is the fluorescence of the zinc over and above a very small background of the flame.

Using nondispersive techniques, sensitivity of the method is greatly increased. Another important feature of the system is the effort taken to collect as much radiation from the atomizer as possible. In atomic absorption no special effort is made to do this, since the only interest is the measure of the degree of absorption of the hollow cathode lamp, and special collection devices are not necessary. A schematic diagram of a nondispersive system is shown in Fig. 3.7.

Other atomizers that have been used include electrothermal atomizer and plasma (10, 11), and the sensitivities obtained using a plasma atomizer are shown in Table 3.1.

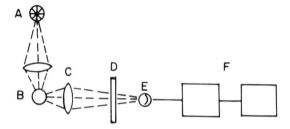

Figure 3.7 Nondispersive atomic fluorescence equipment. (A) Light source; (B) atomizer; (C) gathering lens; (D) filter; (E–F) detector and readout. After West and Cresser (6).

Table 3.2 Plasma/AFS Detection Limits

Element	μg/L	Element	μg/L
Nonrefractory metals			
Ag	2	Mg	0.5
As	400	Mn	3
Au	15	Ni	5
Bi	80	Pb	70
Ca	0.4	Pd	20
Cd	0.5	Pt	60
Co	3	Rh	3
Cr	8	Sb	250
Cu	1	Se	100
Fe	10	Sr	2
Ga	20	Te	200
Hg	400	Tl	20
In	20	Zn	0.4
Alkali metals			
K	0.8	Na	0.3
Li	0.4		
Refractory metals			
Al	20	Si	200
B	500	Sn	200
Ba	5550	Ti	300
Be	2	V	100
Ge	200	W	700
Mo	50	Y	500

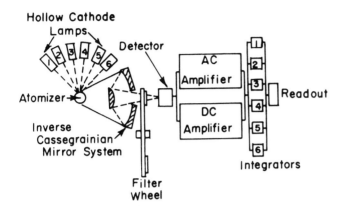

Figure 3.8 Multichannel nondispersive atomic fluorescence spectrometer. After West and Cresser (6).

e. Multielement Analyses

Simultaneous multielement analyses have been achieved in atomic fluorescence using the equipment outlined in Fig. 3.8. In this system a series of light sources, all focused on the atomizers, are used. Radiation from the individual light sources falls on the atomizer in sequence. Fluorescence from the atomizer is focused onto a detector and readout system.

The detectors are programmed to distinguish the fluorescence generated by each separate radiation source. This system is comparatively simple to use and quite inexpensive and offers a convenient way for multielement analysis.

Sophisticated equipment has been reported which uses an ICP as an atomizer for simultaneous multielement analysis. High sensitivity is claimed (12). Further, using an ICP and a pulsed, high-current hollow cathode as an excitation source, not only is high sensitivity claimed, but a linear analytical range of four decades was reported (13). The data were collected for Cu, Ag, Zn, Al, Cr, and Mo atomic lines and Cu, Cr, Zn, and Sr ion lines.

Atomic fluorescence is a very sensitive technique, as was shown in Table 3.2. However it is an elusive technique for quantitative analyses. Most analytical chemists prefer to use more reliable methods for quantitative analysis. However, one interesting application is for the determination of plutonium in nuclear fuel reprocessing (14).

Table 3.3 Comparison of Detection Limits for Atomic Emission (AES), Atomic Absorption (AAS), and Atomic Fluorescence Spectroscopy (AFS)

Element	Wavelength (nm)	Detection limit (ppm)		
		AES[a]	AAS[a,b]	AFS[c]
Ag	328.07	0.008	0.001 (A)	0.0001
Al	396.15	0.05		0.005
	309.28		0.1 (N)	
As	193.70	10	0.03[f]	0.1
Au	267.60	2		0.05
	242.80		0.02 (N)	
B	518.0[a]	0.05		
	249.68		2.5 (N)	
Ba	553.55	0.002	0.02 (N)	
Be	234.86	1	0.002 (N)	0.01
Bi	306.77	20		
	223.06		0.05 (A)	0.05
Ca	422.67	0.0002	0.002 (A)	0.000001
Cd	326.11	0.8		
	228.80		0.001 (A)	0.00001
Ce	569.92	10		0.5
Co	345.35	0.03		
	240.72		0.002 (A)	0.005
Cr	425.43	0.004		
	357.87		0.002 (A)	0.004
Cs	455.53	0.6		
	852.11		0.05 (A)	
Cu	324.75	0.01	0.004 (A)	0.001
Dy	404.60	0.05		
	410.39		0.2 (N)	
Er	400.80	0.07	0.1 (N)	0.5
Eu	459.40	0.0005	0.04 (N)	0.02
Fe	371.99	0.03		
	248.33		0.004 (A)	0.008
Ga	417.21	0.06		0.01
	287.42		0.05 (A)	
Gd	440.19	5		
	622.09[d]	0.07		
	368.41		4 (N)	
Ge	265.12	0.4	0.1 (N)	20
Hf	531.16 (II)[e]	20		
	286.64		20 (N)	
Hg	253.65	10	0.5 (A)	0.002
Ho	410.38	0.1	0.1 (N)	

Table 3.3 (*Continued*)

Element	Wavelength (nm)	Detection limit (ppm)		
		AES[a]	AAS[a,b]	AFS[c]
In	451.13	0.003		0.002
	303.94		0.03 (A)	
Ir	380.01	3		
	550.0[a]	0.4		
	284.97		1 (N)	
K	766.49	0.00005	0.003 (A)	
La	550.13	6	2 (N)	
	441.82[d]	0.01		
Li	670.78	0.00002	0.001 (A)	
Lu	451.86	1		
	331.21		3 (N)	
Mg	285.21	0.07	0.003 (A)	0.001
Mn	403.31	0.008		
	279.48		0.0008 (A)	0.002
Mo	390.30	0.2		
	313.26		0.03 (N)	
Na	589.00	0.0005	0.0008 (A)	
Nb	405.89	1	3 (N)	1
Nd	492.45	0.7		
	463.42		2 (A)	
Ni	352.45	0.02		
	232.00		0.005 (A)	0.003
Os	442.05	2		
	305.87		0.4 (N)	
Pb	405.78	0.1		0.01
	283.31		0.01 (A)	
Pd	363.47	0.05		
	247.64		0.01 (A)	
Pr	495.14	0.07	4 (N)	
Pt	265.94	4	0.05 (A)	
Rb	780.02	0.008	0.005 (A)	
	794.76	3		
Re	346.05	0.2	0.6 (N)	
Rh	343.49	0.03	0.02 (A)	
Ru	372.80	0.3		
	349.89		0.6 (A)	
Sb	252.85	0.6		
	217.58		0.3 (A)	

(*Continued*)

Table 3.3 (*Continued*)

Element	Wavelength (nm)	Detection limit (ppm)		
		AES[a]	AAS[a,b]	AFS[c]
Sc	402.37	0.8		
	391.18		0.1 (N)	
Se	196.03	100	0.1[f]	0.04
Si	251.61	3	0.1 (N)	
Sm	476.03	0.2		
	429.67		0.6 (N)	
Sn	284.00	0.1		
	235.48		0.05 (A)	
Sr	470.73	0.0005	0.005 (A)	0.01
Ta	474.02	4		
	271.47		3 (N)	
Tb	432.65	0.5	2 (N)	
	534.0[d]	0.03		
Te	486.62	2		
	214.28		0.05 (A)	0.05
Th	491.98 (II)[e]	10		
Ti	334.90	0.2		
	364.27		0.1 (N)	
Tl	535.05	0.02		
	276.79		0.02 (A)	
	377.57	0.1		0.008
Tm	371.79	0.08	0.04 (N)	0.1
U	544.8[d]	5		
	351.46		20 (N)	
V	437.92	0.1		
	318.40		0.02 (N)	
W	400.88	0.6	3 (N)	
Y	362.09	1		
	597.2[d]	0.03		
	407.74		0.3 (N)	
Yb	398.80	0.006	0.02 (N)	0.01
Zn	213.86	10	0.001 (A)	0.0002
Zr	360.12	5	4 (N)	

[a]Adapted from G. D. Christian and F. J. Feldman, *Appl. Spectrosc.*, 25: 600 (1971). Nitrous oxide–acetylene flame.

[b]Fuel is acetylene. Letter in parentheses indicates the oxidant. A = air, N = nitrous oxide.

[c]From V. A. Fassel and R. N. Knisely, *Anal. Chem.*, 46: 1110A (1974).

[d]Band emission.

[e]Ion line.

[f]Argon-hydrogen-entrained air flame.

REFERENCES

1. Wood, R. W., *Phil. Trans.*, *10*: 513 (May, 1905).
2. Robinson, J. W., *Anal. Chim. Acta.*, *22*: 254 (1961).
3. Winefordner, J. D., T. J. Vickers, and R. A. Staab, *Anal. Chem.*, *36*: 161 (1964).
4. Winefordner, J. D., *Anal. Chem.*, *36*: 165 (1964).
5. West, T. S., et al., *Talanta*, *13*: 805 (1966).
6. West, T. S., an M. S. Cresser, *Appl. Spec. Rev.*, *7*: 79 (1973).
7. Willis, J. B., *Aust. J. Sci. Res.*, *A4*: 172 (1951).
8. West, T. S., et al., *Talanta*, *14*: 1151 (1967).
9. Perkins, L. D., and G. L. Long, *Applied Spectroscopy*, *42.7*: 1285 (1988).
10. Pereli, F. R., Jr., J. P. Dougherty, and R. G. Michel, *Anal. Chem.*, *59*: 1784 (1987).
11. Demens, D. R., D. A. Busch, and C. D. Allemand, *American Laboratories*: 168 (March, 1982).
12. Demens, D. R., D. A. Busch and C. D. Allemand, American Lab: 167 (March, 1982).
13. Masamba, W. R., B. W. Smith, R. J. Krupta, and J. D. Wineforder, *Appl. Spec.*, *42(5)*: 872 (1988).
14. Bertholed, T., P. Mauchien, A. Vian, and P. L. Prevost, *Applied Spectroscopy*, *41*, (11): 913 (1987).

4

Flame Photometry

When certain metals are put into a flame it becomes brightly colored. The Chinese used sodium and many other metals to form brilliantly colored flames and were the first to produce firework displays based on this knowledge. For at least 100 years these colored flames have been used as a qualitative test for certain metals, particularly the alkaline metals and the alkaline earths. This is useful for qualitative analysis; however, it was very difficult visually to estimate quantitatively the intensity of these colors and therefore deduce the concentration of the metals present.

This problem in quantitation was overcome by the advent of the spectrophotometer. This instrument permitted us to select the wavelengths of the radiation and measure its intensity with considerable accuracy. The spectrophotometric technique has proven to be one of the most reliable and easily used techniques for the determination of concentrations of sodium, potassium, calcium, and magnesium. The concentrations of these four elements are of paramount importance to the metabolism and health of most members of the plant and animal kingdoms, including human beings, because the sodium-to-potassium ratio controls the action of muscles, including the heart, and calcium and magnesium have profound effects on the functioning of the nerves in the body. It is also important in agriculture with respect to the soil because it affects the plants grown in that soil.

Recently however, flame photometry has not been confined to these four elements but has been extended to many other metals in the Periodic Table. These are listed in Tables 4.1 and 4.2. The concentration of these elements may be

Table 4.1 Wavelengths of Emission Lines Used for the Detection and Determination of Some Common Elements, Preferred Flames Used, and the Detection Limits

Element	Wavelength (nm)	Type of flame	Detection limits (ppm)
Aluminum	396.2	OA	0.1
	484.0	OA	0.5
Antimony	252.8	OA	1.0
Arsenic	235.0	OA	2.2
Barium	455.5	OH	3
	553.6	OH	1
Bismuth	223.1	OA	6.4
Boron	249.8	OA	7
	518.0	OA	3
Cadmium	326.1	AH	0.5
Calcium	422.7	OA	0.07
	554.0	OA	0.16
	662.0	OA	0.6
Cesium	455.5	OH	2.0
	852.0	OH	0.5
Chromium	425.4	OA	5.0
Cobalt	242.5	OA	1.7
	353.0	OA	4.0
Copper	324.7	OA	0.6
Gallium	417.2	OA	0.5
Gold	267.6	OA	2.0
Indium	451.1	OH	0.01
Iron	372.0	OA	2.5
	386.0	OA	2.7
	550.0	OA	0.5
Lanthanum	442.0	OA	0.1
	741.0	OA	4.5
Lead	405.8	OA	1.0
Lithium	670.8	OA	0.007
Magnesium	285.2	OA	0.8
	383.0	OA	1.6
Manganese	403.3	OA	0.01
Mercury	253.6	OA	2.5
Molybdenum	379.8	OA	0.5
Neodymium	555.0	OH	0.2
	702.0	OH	1.0
Nickel	352.4	OA	0.1
Niobium	405.9	OA	12
Palladium	363.5	OH	0.1

Table 4.1 (*Continued*)

Element	Wavelength (nm)	Type of flame	Detection limits (ppm)
Phosphorus	253.0	OH	1.0
Platinum	265.9	OA	10
Potassium	404.4	OH	1.0
	767.0	OH	0.01
Rhenium	346.1	OA	0.3
Rhodium	369.2	OH	0.1
Rubidium	780.0	OH	0.3
Ruthenium	372.8	OA	0.3
Scandium	604.0	OH	0.012
Silicon	251.6	OH	4.0
Silver	328.0	OH	0.1
	338.3	OH	0.6
Sodium	590.0	OH	0.001
Strontium	460.7	OA	0.01
Tellurium	238.6	OA	2.0
Thallium	377.6	OH	0.6
Tin	243.0	OA	0.5
Titanium	399.9	OA	1.0
Vanadium	437.9	OA	1.0
Yttrium	597.0	OA	0.3
Zinc	213.9	OA	77

Key: AH means air–hydrogen flame; OA, oxygen–acetylene flame; OH, oxygen–hydrogen flame.

measured by flame photometry, but frequently that technique is not the method of choice because of the relatively low energy available from a flame and therefore the relatively low intensity of the radiation from the metal atoms, particularly those that require large amounts of energy to become excited. Although flame photometry is a means of determining the total metal concentration of a sample, it tells little or nothing about the molecular form of that metal in the original sample. It has not been used for the direct detection and determination of the noble metals, halides, or inert gases. All of these elements require more energy than the flame provides in order to become excited.

ORIGIN OF SPECTRA

An atom is composed of a nucleus and orbiting electrons. The size and shape of the orbitals in which the electrons travel around the nuclei are not random. For

Table 4.2 Sensitivity Results Obtained with Nitrous Oxide–Acetylene Flames

	Sensitivity (ppm) obtained by		
Element	Fisher Scientific Co.[a]	Fassel[b]	Picket and Koirtyohann[c]
Al	0.1	0.03	0.005
Ba	0.08	0.002	
B	100		
Cd		2.0	
Ca	0.005	0.001	
Cr	0.02	0.02	
Co	0.5		
Cu	0.1	0.05	
Dy	0.5	0.07	
Eu	0.1	0.0006	
Gd	1.0	2.0	
Ga			0.2
Ge	4.0		0.5
In	0.1	0.03	0.002
Fe	1.0	0.7	
Pb	3.0	3.0	
Mg	0.05	0.2	
Mn	0.7	0.1	
Hg	100	40	
Mo	1.0	0.2	
Ni	5.0	0.6	
P	10.0		
K	0.02	0.003	
Na	0.001	0.0001	
Sr	0.02	0.004	
Th			0.02
Ti	0.3	0.4	
V		0.07	
Zr	0.5		

[a]Data courtesy of Fisher Scientific Company, Waltham, Mass.
[b]Data from V. A. Fassel, *Spectrochim. Acta, 24B:* 1494 (1969).
[c]Data from E. Picket and S. R. Koirtyohann, *Spectrochim. Acta, 24B:* 325 (1969).

each orbital only certain permitted energy levels are permissible as described mathematically by the quantum theory. This has been discussed in Chapter 1.

In the case of the atom, the valence electron is the most easily excited. If the valence electron is orbiting in the lowest permitted energy state, the atom is said to be in the ground state. When an excited atom is put into a flame, the valence electron becomes thermally excited for a short period, emits radiation, and returns ultimately to the ground state. The population in the excited state is governed by the Boltzmann distribution (see page 11). The measurement of the wavelengths and intensity of this emitted radiation is the basis of flame photometry.

The permitted energy levels are the empty orbitals of that particular atom as described in Chapter 1. These are illustrated in Fig. 4.1, which is a partial diagram of the energy levels of lithium. This is called a Grotrian diagram. A transition of an electron from one energy level, E_2, to another, E_1, involves the loss of well-defined energy E_2 and E_1.

But as we have seen,

$$E_2 - E_1 = h\upsilon$$

where

h = Planck's constant
υ = frequency of the emitted light

since

$$\nu = \frac{c}{\lambda}$$

where

c = speed of light
λ = wavelength of the light

$$E_2 - E_1 = \frac{hc}{\lambda}$$

or

$$\lambda = \frac{hc}{E_2 - E_1}$$

The energy levels E_2 and E_1 are characteristic of the emitting element, and therefore the wavelength λ of the radiation emitted is also characteristic of the particular element. It is a physical property of that element which does not change under different conditions. Consequently when flame photometry is used as an analytical tool, the wavelengths of the radiation emitted from the flame indicate to us which elements are causing the radiation. Also, the intensity of the radiation is related to the concentration of those emitting elements.

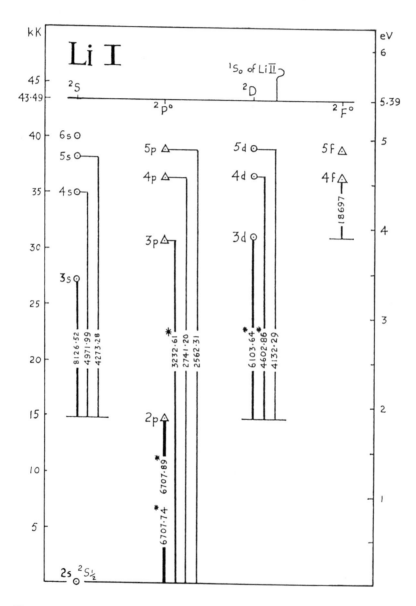

Figure 4.1 Level diagram of lithium. (From Ref. 5)

1. Energy of the Flame

The energy available in the flame for exciting atoms is dependent on the flame temperature as defined by the Boltzmann distribution. Flames have only limited amounts of energy available when compared to excitation sources such as emission spectrographs or plasma emission torches. For this reason, flames are most useful only for elements that require low amounts of energy to become excited. This means that the method is most useful for the determination of the alkali metals and alkaline earth metals. They are not the method of choice for transition metals and most of the other metals in the Periodic Table. These metals generally require significantly more energy in order to become excited. They are most often done by emission spectrography or plasma emission where much higher energy is available. However, it must also be stated that when we try to analyze Group I and Group II elements in these high energy sources, we run the risk of not only exciting them but also ionizing them. Although the Boltzmann distribution would lead us to believe that the emission intensity would be greatly increased in plasma emission for the Group I and Group II elements, in practice it is found not to be so because the atoms are ionized and this results in a loss of atoms and the generation of ions. Ionization causes a complete change of energy levels and, therefore, the entire emission spectrum, with a reduction in the intensity of the atomic spectra.

The flame is a source of intense radiation especially for atoms inside the flame. It is probable that this is a source of excitation of the atoms (radiation excitation) which augments the thermal excitation expected from the Boltzmann distribution. This point has not been extensively studied but provided a catalyst for the evolution of atomic fluorescence (1).

For example, a fuel-rich oxyacetylene flame generates intense radiation bands at short wavelengths (300–200 nm). Iron solutions injected into these flames strongly emit iron lines at many wavelengths down to 200 nm. On the other hand, oxygen-rich flames operating at the same temperature do not emit such broad-band radiation, and iron lines are not emitted from such flames (1). The emission intensity generated by thermally excited atoms should be the same according to the Boltzmann distribution but is clearly not the same. In the same paper, a description of atomic fluorescence was described using a hollow cathode as a light source (1).

The advent of the nitrous oxide acetylene flame has provided us with a flame with a higher temperature than was hitherto available. This has increased the usefulness of flame photometry. Also, the use of the elongated burner as developed for atomic absorption spectroscopy has increased the sensitivity of the method. The function of the flame is discussed in greater detail on page 91.

B. EQUIPMENT

A schematic diagram of the equipment used in flame photometry is shown in Fig. 4.2. The instrument has the same basic components as the spectroscopic equip-

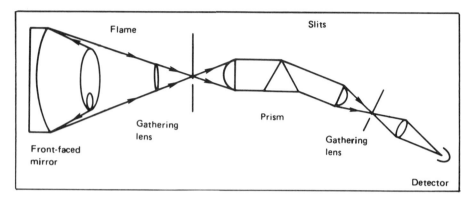

Figure 4.2   Schematic diagram of a flame photometer.

ment, namely the source (the flame), a monochromator to select wavelengths, and a detector to record the radiation intensity. The flame is generated by a burner, which atomizes and simultaneously excites the atoms of the sample. The various components of the instruments are described below.

1. Burner

The central component of a flame photometer is the burner. It has two functions: first, it vaporizes the sample and introduces it into the flame, where free atoms are formed; second, the flame excites the atoms and causes them to emit radiation. The intensity of emission is dependent on the number of atoms in the excited state, which is controlled by the Boltzmann equation. In addition, it is possible for atoms to become excited by the radiant energy in the flame. There are two types of burners: the total consumption burner and the Lundergardh burner, the latter being much more common in modern equipment.

In the total consumption burner, the sample must be in the form of a liquid, evaporated completely into the flame. A typical total consumption burner is shown in Fig. 4.3. In this burner the air or oxygen aspirates the sample into the base of the flame. Evaporation, atomization, and excitation of the sample then follow. Atomization means the reduction to the free atomic state, not reduction to a small particle size. The atomization step in flame photometry is exactly the same as that encountered in atomic absorption spectroscopy.

In the Lundegardh burner (Fig. 4.4), the sample must be in liquid form. It is aspirated into the spray chamber. Large droplets condense on the side and drain away; small droplets and vaporized sample are swept into the base of the flame

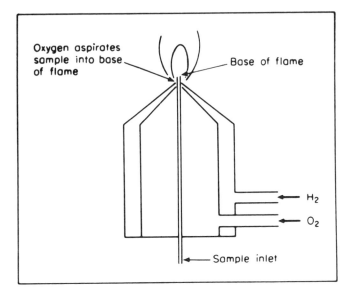

Figure 4.3   Beckman total consumption burner.

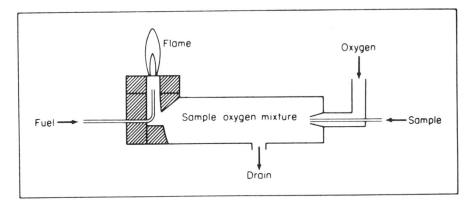

Figure 4.4   Lundegardh burner.

in the form of a cloud. Various devices have been used to enhance the nebulization stage in this type of burner. These include the use of the impact bead (Perkin-Elmer), ultrasonic vibrators (2), and, more recently, thermospray heaters (3).

An important feature of this burner is that only about 5% of the sample reaches the flame. The rest of the droplets condense and are drained away. This is a significant loss in atomization efficiency and, therefore, sensitivity.

Difficulties may also arise if there is any selective evaporation of the solvent in the spray chamber. In particular, if there are two solvents, the more volatile will preferentially evaporate, leaving the element of interest in the less volatile component. The latter may drain away, taking the sample with it. In this event many of the sample atoms never reach the flame, and the emission intensity is reduced and an incorrect analysis obtained.

An advantage of the Lundegardh burner is that it is physically quiet to operate, which is a distinct advantage over the noisy total consumption burner.

The efficiency of the Lundegardh has been improved by putting an impact bead immediately before the nebulizer (see Fig. 4.5). This impact bead has the effect of breaking up large droplets as they emerge from the nebulizer, thereby increasing the number of small droplets and consequently the efficiency of nebulization. The atomization step is very dependent on the efficiency of the nebulization since the droplets must be evaporated and the residue broken down to liberate free atoms after introduction into the flame (see page 97). A number of factors affect the drop size, d_0, including viscosity η, density ρ, surface tension γ of the sample solution, gas and the quantity aspirated solution ϕaq liquid, and the velocity of the nebulizing gas. These variables are related in the empirical expression:

$$d_0 = \frac{585}{v} \left(\frac{\gamma}{\rho}\right)^{0.5} + 597 \left(\frac{\eta}{(\gamma\rho)^{0.5}}\right)^{0.45} 1000 \left(\frac{\phi \text{ liq}}{\phi \text{ gas}}\right)^{1.5} \qquad (4.1)$$

This diameter d_0 controls the volume-to-surface relationship, or *Sauter mean*. It can be seen from this expression that it is important to maintain a constant viscosity, surface tension, density, and flow rates of the pertinent gases to reproduce the nebulized drop size and therefore the emission atomization process intensity. To that end it is essential that the standards and sample be analyzed under identical flame conditions and that the solvents and matrices be as close to each other as possible.

a. Shielded Burners

T. S. West et al. developed shielded burners in which the flame (particularly the reaction zone) was shielded from the ambient atmosphere by a stream of inert gas. This shielding leads to a quieter flame and better analytical sensitivity. Table 4.3 shows results obtained with commercial equipment based on this technique and developed by the Beckman Instrument Co. (Fullerton, CA).

Table 4.3 Results Obtained with Shielded Burners[a]

Element	Sensitivity (ppm)
Ba	0.05
Bi	2
Ca	0.002
Cr	0.0007
Co	0.04
Pb	0.05
Mg	0.3
V	10

[a]Data courtesy of Beckman Instrument Co., Fullerton, Calif.

2. Mirrors

The radiation from the flame is emitted in all directions in space. Much of the radiation is lost, and loss of signal results. In order to maximize the amount of radiation used in the analysis, a mirror is located behind the burner to reflect the radiation back to the entrance slit of the monochromator. This mirror is concave and covers as wide a solid angle from the flame as possible. To get the best results, the hottest and steadiest part of the flame is reflected onto the entrance slit of the monochromator. This helps reduce flame flicker from upper parts of the flame where light intensity is reduced and noise is increased.

The reflecting surface of the mirror is *front-faced*. If the reflecting surface were on the rear as in the normal household mirror, the radiation from the flame would have to go through the support material such as glass or quartz twice before it is reflected to the entrance slit. Since the support material absorbs some radiation, there would be a considerable loss of signal, particularly at the shorter wavelengths. Front surface mirrors are most efficient, but they are not physically protected. They are very easily scratched and subject to chemical attack, e.g., from acid vapors in a hood. Great care should be taken to protect them by keeping the instrument away from corrosive atmospheres.

3. Monochromator System

The monochromator consists of a dispersion element and a set of slits. The dispersion element separates the radiation according to wavelengths. The common dispersion elements used are prisms or gratings.

The function of the entrance slit is to prevent stray radiation from entering the light path, allowing the radiation from the flame and that reflected from the mirror to proceed to the dispersion element. The function of the exit slit is to select the wavelengths to be monitored for measurement. The monochromator system is described more fully on page 66.

a. The Use of Filters

An advantage of flame photometry is that the flame is of fairly low energy and therefore only a few spectral lines are emitted. This results in a simple spectrum, and high resolution usually is not necessary for most analysis. In highly routine analysis a filter is used as the monochromator. The filters are made of materials which are transparent to a small wavelength range. That wavelength range is selective to the emission from the element of interest. When the filter is placed between the flame and the detector, radiation of the pertinent wavelength reaches the detector and can be measured. Other radiation is absorbed by the filter and does not contribute to the signal.

Filters have been designed for use in the determination of sodium, potassium, calcium, magnesium, and a few other elements. They are particularly useful in clinical laboratories where many analyses of these four elements are performed daily. Of course, the instrument is seldom used for any other elements, but in a fully routine laboratory this is not a handicap. There is a considerable decrease in cost of the instrument.

4. Detectors

The function of the detector is to measure the intensity of radiation falling on it. The most common detector is the photomultiplier, which generates an electrical signal from the radiation falling on it. In less expensive instruments, barrier layer cells are also used. Photomultipliers and barrier layer cells are described in Chapter 2, Sec. C.9.

5. Flames

When atoms are generated and excited in flames, the thermal excitation is not great, and only the lower excitations states are populated. (The complicated process and the factors that affect atomization are shown in Table 2.3.)

There are many factors that affect atomization efficiency. These include the viscosity of the solvent, the surface tension of the solvent, the predominant anion, the flame temperature, and the composition of the flame, whether it is a reducing flame, stoichiometric, or an oxidizing flame. Finally, the stability of the free atoms dictates their lifetime, because they are in the highly reactive environment of the flame. Free atoms rapidly become oxidized, and if the oxide is stable at the flame

Table 4.4 Flame Temperatures of Typical Flames

Fuel	Oxidant	Flame temperature (°C)
H_2	O_2	2800
H_2	Air	2100
H_2	Ar	1600
Acetylene	O_2	3000
Acetylene	Air	2200
Acetylene	N_2O	3000
Propane	O_2	2800
Propane	Air	1900

temperature, it stays oxidized and no longer contributes to the analytical signal. However, if it is thermally decomposed (e.g., $Na_2O \rightarrow 2Na + O$), the atom population and therefore the light intensity remains high.

a. Flame Temperature

The flame temperature depends very much on the fuel and the oxidant used. The most common fuels are acetylene, hydrogen, propane, and "town gas." The most common oxidants are air, oxygen, and nitrous oxide. The temperatures relevant to these flames are listed in Table 4.4.

The flame temperature affects a number of steps in the atomization process. It affects the number of atoms excited directly as shown by the Boltzmann distribution. Second, the decomposition of the residue is affected by the flame temperature. The higher the flame temperature, the faster the decomposition. The same chemical interferences occur in flame photometry. Hotter flames reduce chemical interferences because there is an increase in the decomposition rate and therefore an increase in the atomization efficiency (see page 96.)

It should be stated at this point that the optimum experimental conditions for measuring the concentration of various elements has been worked out by research workers in the field and are often provided by the instrument manufacturers.

b. Use of Nitrous Oxide–Acetylene Flames

During research in atomic absorption spectroscopy, J. Willis and M. Amos independently found that nitrous oxide–acetylene flames were superior to other flames for efficiently producing free atoms. This was particularly true for metals with very reflective oxides, such as aluminum and titanium. Later, workers in the field of flame photometry found that this same flame was very useful in flame photometry (see Table 4.2). However, the high temperature reduces its usefulness

Table 4.5   Percent Ionization in Flames

Element	Ionization potential (eV)	Acetylene/ Air (2400°C)	Acetylene/ Oxygen (3100°C)	Acetylene/ Nitrous oxide (3200°C)
Lithium	5.39	0.01	16	—
Sodium	5.14	1.1	26	—
Potassium	4.34	9.0	82	—
Rubidium	4.18	14.0	89	—
Cesium	3.89	29.0	96	—
Magnesium	7.64		0.01	6
Calcium	6.11	0.01	7	43
Strontium	5.69	0.01	17	84
Barium	5.21	2	42	88
Manganese	7.43			5

*Based on Atomic Absorption Spectroscopy by W. Slavin, 1968 p.69, Interscience New York.

for the determination of the alkali metals because they are easily ionized, as shown in Table 4.5.

One problem encountered with this type of flame was the intense background emission, which makes measurement of the metal emission very difficult. However, the "wobbler" designed by Rains for background correction and the commercial nitrous oxide–acetylene burners developed provide equipment capable of high sensitivity and accuracy. Results obtained by Fisher Scientific using this technique are shown in Fig. 4.5. Note the linear relationship between the emission intensity and the concentration of the metal. Note also the extended useful analytical range of the method. The calibration data are drawn to log-log paper for convenience.

c. Solvents

The solvent used in the sample affects the signal in two ways. First, viscosity controls the rate at which the sample is aspirated into the flame. There is an optimum sample flow rate, which has to be experimentally determined. If the flow rate is too great, the flame is swamped and the signal drops off; if the flow rate is too low, then the signal is decreased because insufficient sample finds its way into the flame.

The second effect of the solvent is caused by the difference between aqueous or organic materials. If it is aqueous, then the sample requires energy to evaporate it. Generally an inorganic salt is left, which requires more energy from the flame to decompose it. These are two endothermic steps, which slows down the atomization process. On the other hand, if the solvent is organic, it burns on introduc-

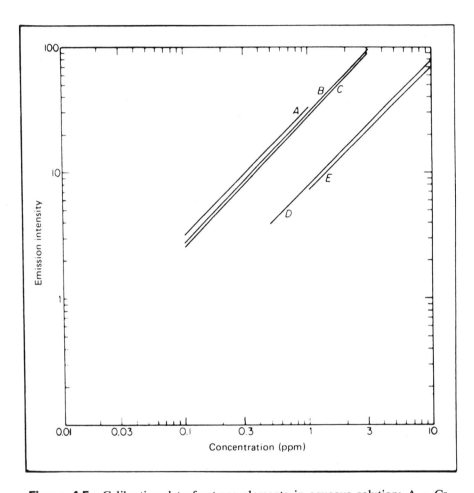

Figure 4.5 Calibration data for trace elements in aqueous solution: A = Cr 475.4, N_2O–C_2H_2; B = V 440.8, N_2O–C_2H_2; C = Al 396.1, N_2O–C_2H_2; D = Mn 403.1, N_2O–C_2H_2; E = Ti 365.3, N_2O–C_2H_2. (Courtesy of Fisher Scientific Co., Boston, Mass.)

tion to the flame and usually leaves an organic residue, which in turn burns inside the flame. Each of these steps is exothermic, the atomization efficiency is increased, and there is an enhancement of signal. This is the reason for signal enhancement when organic solvents are used rather than aqueous solvents.

It can be seen from these comments that in order to get good quantitative data it is necessary that flame conditions, including the fuel, oxidant, and solvent used, should be the same for both the samples to be analyzed and used in preparing calibration curves. Any variations in these conditions cause a major variation in flame intensity and therefore in the analytical data obtained.

C. FLAME EMISSION

1. Emission of Atomic Spectra

The intensity of atomic spectra depends on the number of atoms in the excited state. This is related directly to the temperature of the system and is given by the expression

$$S = \frac{N_1}{\tau} = \frac{N_0}{\tau} \frac{g_1}{g_2} e^{-E/kT} \tag{4.2}$$

where

S = intensity of emission line
N_1 = number of excited atoms
N_0 = number of unexcited atoms
τ = lifetime of the atoms in the excited state
E = energy of the excitation
$\frac{g_1}{g_2}$ = ratio of statistical weights of the ground state to the excited state
T = absolute temperature
k = Boltzmann distribution coefficient

This relationship has been described on page 22.

There are two components that must be strictly controlled in order to get reproducible data. One is T, the absolute temperature, which is controlled by the flame composition. The second is N_1, the number of excited atoms. This is related directly to the number of unexcited atoms by the equation shown above (Eq. 4.2). The number of unexcited atoms is related to the concentration of the element in the original sample and to the atomization efficiency of the flame. It is vital that the atomization efficiency be kept rigidly controlled if reproducible data are to be obtained.

The energy of excitation can be deduced from the Grotrian diagrams. The excited energy state most easily populated is the first excited state. For some elements the amount of energy required is very high. These include elements such as zinc, cadmium, mercury, and arsenic. Consequently, flame photometry is note useful in the determining of these elements. On the other hand, elements such as sodium, potassium, lithium, caesium, calcium, strontium, and barium are easily excited, and the flame is one of the most attractive methods for their determination.

2. Emission of Spectra by Molecules

In addition to the atomic spectra emitted by elements in the flame, we can also detect molecular spectra from excited molecules such as metallic oxides and somecarbon compounds. These are usually band spectra and therefore not as intense as line spectra. They are, however, analytically useful, particularly for higher concentrations. Examples of these fragments are barium hydroxide, calcium hydroxide, tin hydroxide, and manganese hydroxide along with calcium oxide and the oxides of the rare earth. All emit molecular band spectra when introduced to the flame and can be used for the determination of barium, calcium, maganese, and rare earth, respectively.

Organic solvents also emit strongly during the process of combustion [4]. The emission spectra are strong but depend very much on flame conditions. If the flame is reducing, then the organic solvents are incompletely burned and the free radicals and other fragments emit strongly at characteristic wavelengths. Howver, if the flame is oxidizing, then it is probable that the organic solvent will be completely burned, producing CO_2 and water. Emission spectra are then observed for these compounds and fragments of these compounds, such as OH bands. At intermediate flame condition, flame fragments and CO_2 and water are observed.

This should be remembered when checking the background radiation from a flame. It can be intense and variable depending on flame conditions. For standards and samples to relate to each other, flame conditions must be kept constant.

3. Background Emission

The flame is a source of spectral energy arising from combustion of the fuel used to create it. Furthermore, the hot or burning solvent and other components of the sample also emit radiation over an extensive wavelength region. These two sources of radiation form the background radiation that is always present in the spectrum of the flame. If necessary, a correction must be applied to quantitative measurements of emission line intensities. A typical example of background emissions from an oxygen–acetylene flame and N_2O–acetylene are illustrated in Fig. 4.6. The principal emission lines of nickel, sodium, and potassium are superimposed on this background for illustrative purposes. The background can be measured either from a blank at the emission line wavelength or by measuring the background of the actual sample very close to the emission line. The net emission from the sample is the total emission minus the background.

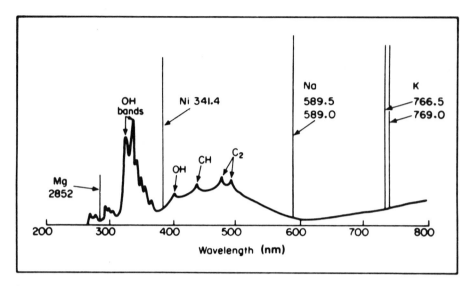

(a)

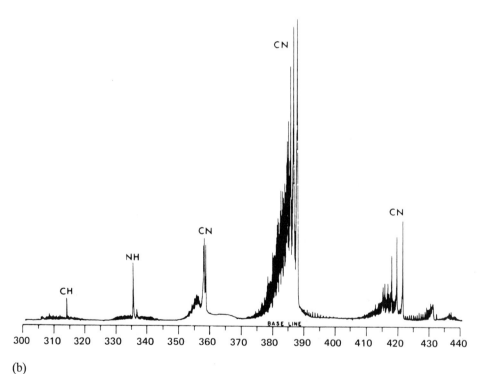

(b)

Figure 4.6 (a) Total emission spectrum of an oxyacetylene flame. (b) Total emission spectrum of nitrous oxide–acetylene.

T. Rains at the U.S. National Bureau of Standards developed an ingenious method to overcome this problem. He designed a reflector plate that oscillated at a controlled frequency and through a controlled angle. At one end of the oscillation of the reflector plate the emission line from the metal falls onto the detector; at the other end of the swing the background emission falls on the detector. As a result, the detector is exposed alternately to the background and to the metal emission. It therefore generates an AC signal that is proportional to the difference between the two signals. Effectively, correction is made for the background emission, and the intensity of the emission line is measured more accurately.

D. ANALYTICAL APPLICATION

1. Qualitative Analysis

The qualitative aspects of flame photometry are useful mostly for the detection of elements in Groups I and II of the Periodic Table. These elements include sodium, potassium, lithium, magnesium, calcium, strontium, and barium. The presence of certain elements can be detected visually, as in the case of the yellow flame produced by sodium. It is generally much safer, however, to use a filter or a monochromator to separate radiation with the wavelengths characteristic of the different metals from other radiation present. The yellow radiation from sodium impurities in a sample is often intense enough to mask radiation from other elements present. If radiation of the characteristic wavelength is detected, this is taken to indicate the presence of the corresponding metal in the sample. The method is not as reliable as emission spectroscopy, where radiation at several wavelengths can be examined to confirm the presence of the pertinent element, but it is fast, simple, and, if carried out with care, quite reliable. No information about the molecular structure of the metal compound can be obtained. Further, non-radiating elements, such as carbon, hydrogen, and halides, cannot be detected except under special circumstances. For example, halides can be precipitated using an excess of a standard silver nitrate solution. The excess silver can be determined by flame photometry. From the results, the halide content can be calculated. The method is an indirect determination of the halide, although no radiation from the halide was detected. A list of useful emission lines for the qualitative analysis of the various metals is given in Table 4.1.

2. Quantitative Analysis

The intensity of an emission line is governed by the Boltzmann distribution equation.

Some calculations of the populations of the excited states are shown in Table 1.8. At 2000 K, only one zinc atom in 10^{14} is excited, and even when the temperature increases to 3000 K, it is only one in 10^9 atoms. On the other hand, for sodium at 2000 K the number of excited atoms is one in 10^4, and at 3000 K it increases to one in 10^3. We can see that there are great differences between the population in the excited state depending on the energy required to excite the atoms. This limits the elements that can be usefully excited in flames and enables us to predict which elements can be analyzed by flame photometry and which elements cannot.

It should be pointed out that in the case of zinc the line quoted is at 213.9 nm. This is a resonance line that originates in the ground state. There are many other zinc lines of longer wavelengths, such as the emission line at 636.2 nm, which shows up in emission spectrography. However, this transition is between two excited states and does not include the ground state. Considerably more energy is necessary to populate the higher of these two states than to excite them to the lower excitation state associated with 213.9 nm transition. Therefore, the intensity of such lines is very weak in flame photometry.

It must also be pointed out that we would expect sodium emission to be stronger at the temperature of the higher flame (3000 K). However, in practice this is not so because the sodium atoms not only become excited but ionize, and the number of atoms in the excited state is therefore diminished.

The actual conditions for carrying quantitative analysis using flame photometry have been determined experimentally and have been well documented.

a. Calibration Methods for Quantitative Analysis

To perform quantitative analysis, the sample is introduced into the flame and the intensity of radiation is measured at the pertinent wavelength. The concentration of the emitting metal in the sample is then calculated by one of two methods: (1) the use of calibration curves or (2) the standard addition method. The former method is more commonly used, but under laboratory conditions it is often necessary to use the standard addition method if a calibration curve is not available and there is no time to prepare one or if the matrix of the sample is unknown and the calibration curve is not suitable.

Calibration curves are prepared by making up standard solutions of known concentrations similar to those expected in the sample. For example, if we are to determine on a routine basis samples containing lithium and their normal concentrations are 6 parts per million (ppm), a calibration curve would be prepared using standards ranging from 1 to 20 ppm or a similar suitable range. The actual standard solutions prepared may contain 0, 1, 2, 4, 6, 8, 10, 12, 14, 16, 18, and 20 ppm each. The sample is then analyzed and the intensity of emission from each standard is measured and plotted on a curve. This curve relating emission intensity and the concentration of lithium is the calibration curve. When a sample is run, the emission intensity of the sample is measured, and from the calibration curve the corresponding concentration in the sample can be calculated as shown in Fig. 4.7.

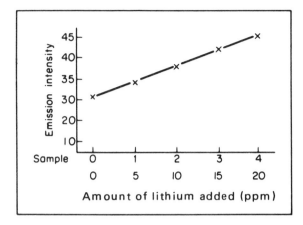

Figure 4.7 Plot of emission intensity against the quantity of lithium added to each aliquot, showing how the standard addition method is used in photometric quantitative analysis.

Suppose, for example, that the emission intensity of the sample was 37. From the calibration curve it can be shown that this is equivalent to approximately 11 ppm.

Note that an emission signal was detected even when no lithium was added to the standard. This is called the background or blank emission signal. It must be corrected for in the final calculation. Also, the relationship between emission intensity and the lithium concentration is linear at low concentration but deviates from linearity at higher concentrations. This is quite common for flame photometric calibration curves. The determination of lithium in a different sample using standard addition is described below.

In the standard addition method, the sample is split into several aliquots. One aliquot is left untreated. To the other aliquots known amounts of the test element are added. For example, the test element may again be lithium. The intensity is then plotted against the quantity of lithium added to each aliquot, as shown in Fig. 4.8.

From Fig. 4.8 the intensity of lithium emission from the original untreated sample was 30 units. When 5 ppm of Li was added to the sample, the intensity increased to 33 units. The addition of 5 ppm of lithium produced an increase in the emission signal of 3 units. Since the emission intensity from the original sample was 30 units, it must have contained $30 \times (5/3)$ ppm of lithium. This procedure is called the standard addition method. One of its most important advantages is that it can be used for samples, the matrix of which is unknown. Any

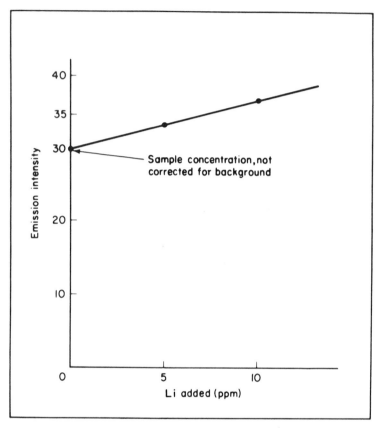

Figure 4.8   Determination of sample concentration from standard addition data.

interference should affect the lithium originally presented and the added lithium in the same manner, and the interference is thus compensated for. A correction should be made for the background emission from the flame. A background emission is produced that is not generated by the lithium present. The intensity of the background emission should be subtracted from the total emission from the sample before the calculations of the lithium content are made. Another advantage of the standard addition method is that it can be used for samples that are rarely analyzed and that would not justify an analytical program to develop a new procedure.

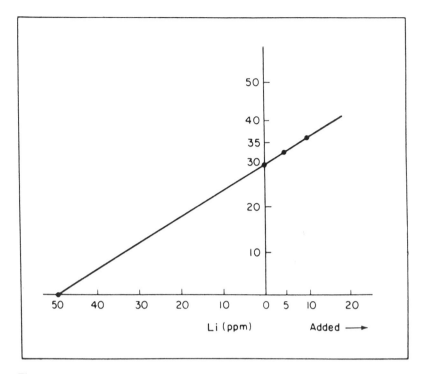

Figure 4.9 Alternative method of calculating the concentration using standard addition. The curve is extended backward to the point of intersection and the concentration read off the axis.

An alternative method of treating the data is to extend the baseline to the left of the vertical axis and extend the line joining the data points until it intersects the baseline. At the point of intersection, the concentration of the sample can be read off. It must be corrected for background. This is illustrated in Fig. 4.9.

3. Interferences

The radiation intensity may not accurately represent the sample concentration because of the presence of other materials in the sample. These materials may cause interference in the analytical procedure. Following are the three principal sources of interference encountered in flame photometry.

a. Radiation Interferences

Two elements or compounds have different spectra, but their spectra may partially overlap and they may both emit at some particular wavelength. The detector cannot distinguish between the sources of radiation and reads out the total signal. If the emission wavelength of the interferent coincides with the wavelength used to measure the radiation intensity from the sample element, an incorrect answer will be obtained. This is a direct source of error. Sometimes it may be corrected by eliminating the effect of the interfering element (e.g., by extraction of the element from the sample) or using calibration curves prepared from a solution containing similar quantities of the interfering material which become part of the calibration signal. However, if the concentration of the interferent is not known, this latter process will not work.

b. Chemical Interferences

The emission intensity depends on how many excited atoms are produced in a flame. If the sample element is in the presence of anions with which it combines strongly, it will not decompose easily. But if the predominant anion combines weakly with the sample element, decomposition may be easier. For example, a given concentration of barium sulfate will give a lower signal than the same concentration of barium chloride because barium chloride is broken down more easily than barium sulfate. This effect is called chemical interference. It is brought about by differences in the strengths of the chemical bonds between the metal and different prominent anions. Any change in an anion can result in chemical interference. The effect can be eliminated by extraction of the anion or by using calibration curves prepared with the same predominant anion at the same concentration as that found in the sample.

c. Excitation Interferences

When an atom is liberated in a flame, it reaches an excited state because of the high temperature of its surroundings. The population of the excited states is given by the Boltzmann distribution. In addition, a number of atoms may be ionized and generate ionic spectra. These neutral, excited, and ionized atoms are all in a state of dynamic equilibrium. The radiation emission intensity becomes steady and provides the basis for flame photometry. If atoms of another species are also present in the flame, they may affect the equilibrium. For example, they may absorb some of the thermal or radiant energy of the flame, thereby reducing the amount of energy available to the sample and increasing their own emission intensity. This in turn will affect the ionic population of Group I and II elements and therefore the intensity of both the atomic emission. It is called *excitation interference*. It has been reported to be particularly severe if the resonance and emission lines of the two elements overlap.

This type of interference is restricted to elements of the first group of the Periodic Table. It can be corrected by preparing calibration curves from standards that contain the interfering element in concentrations similar to that of the sample not always an easy task.

E. CONCLUSIONS

The most useful application of flame photometry is for the rapid quantitative determination of the elements in Groups I and II of the Periodic Table. Using equipment with high optical resolution, other metallic elements may also be determined. Table 4.1 is a list of elements with their emission wavelengths and detection limits.

The principal analytical advantages of flame photometry include its simplicity, sensitivity, and speed. There is, however, a limitation to the number of elements that can be determined by this method. Also, only liquid samples may be used. Sometimes lengthy sample preparation steps are necessary.

Flame photometry gives no information on molecular analysis, but it is used widely for elemental analysis. In soil samples the elements sodium, potassium, calcium, and magnesium are frequently determined. These same elements are determined in plant analysis.

In the medical field, Na, K, Ca, and Mg are routinely determined in blood analysis by flame photometry. These four elements are also determined in urine and other excretions, as well as in body tissue.

In metallurgy sodium is determined in aluminum and aluminum alloys, as well as in metallic calcium. Other examples include the determination of silver in blister copper; magnesium in slags; iron, copper, and cobalt in cobalt mattes; and calcium in cement. Boron has also been determined in various types of organic compounds.

In summary, flame photometry is a simple, rapid method for the routine determination of elements that are easily excited.

REFERENCES

1. Robinson, J. W. *Anal. Chim Acta, 24:* 254 (1961).
2. Robinson, J. W., and J. C. Wu, *Spectroscopy Letters, 18(6):* 399 (1985).
3. Robinson, J. W., and D. S. Choi, *Spectroscopy Letters, 20(4):* 375 (1987).
4. Robinson, J. W., and V. Smith, *Anal. Chim. Acta, 36:* 489 (1966).
5. Chandler, C., *Atomic Spectra,* Van Nostrand, 1964, p. 369.

SUGGESTED READINGS

1. Gaydon, A. G., *The Spectroscopy of Flames,* Wiley, New York, 1957.
2. Gaydon, A. G., and H. G. Wolfhand, *Flames, Their Structure, Radiation and Temperature,* 2nd Ed., Chapman and Hall, London, 1960.
3. Dean, John A., *Flame Photometry,* McGraw-Hill, New York, 1960.
4. Herrman, R., and C. T. J. Alkemade, *Flame Photometry,* trans. P. T. Gilbert, Interscience, New York, 1963.

5

Emission Spectroscopy

A. HISTORY

In 1817 Fraunhoffer noted that when radiation from the electrical discharge between two metal electrodes was passed through a prism it produced a spectrum with lines in it. In 1835 Wheatstone observed that these lines were characteristic of the metals used in the electrodes. Each of these observations was visual since they were made before the advent of photographic plates, which were developed in 1872.

In 1831 Faraday first developed the principle of electrical induction. This was improved by Ruhmkorff in 1850, who developed the induction coil and was able to generate high voltages. Richey in 1857 improved the Ruhmkorff induction coil and was able to produce long sparks up to 12–16 inches in length.

In 1861 Kirchoff used a combination of resistance and induction coils with a Leyden jar to produce sparks across a spark gap. In 1872 Lockyer focused the image of the spark discharge onto the slit of the spectroscope. With this system, he and his coworkers photographed the emission spectrum and were able to study the composition of alloys. In 1874 he stated that *while the qualitative spectrum analysis depends on the position of the lines, the quantitive analysis depends on their length, brightness, thickness, and number.* This provided the foundation of quantitative emission spectrography. In 1879 Lockyer further developed his system by using a DC generator developed by Siemens in Germany.

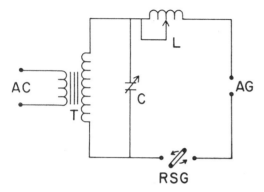

Figure 5.1 Feussner synchronous rotary-gap-controlled high-voltage spark excitation source. T, high voltage transformer; C, variable capacitance; L, variable inductance; RSG, rotary synchronous gap; AG, analytical gap.

Spark discharges were first used to obtain spectra from solution by Lecoq in 1873. Hartley was the first to use graphite electrodes in his studies in 1882.

Various spark sources were developed first with static electrical sources and then with high voltage, low current oscillatory discharges with voltages of between 10,000 and 50,000 V. Fuessner in 1832 first introduced a rotatory interruption spark gap, which provided a more reducible spark discharge. This is shown in Figure 5.1. This circuitry was in common use for many years. It had many advantages in that the inductance could be varied together with the compacitance, thus spark discharges could be controlled depending on conditions required.

By using a number of batteries a DC arc was first used in 1870. The arc gave very intense and therefore sensitive spectra, but the reproducibility was poor. The electronic temperature was in the order of 6,000–8,000 K, and under these conditions metal electrodes melted and vaporized directly. This was an advantage since many metals and alloys can be analyzed directly avoiding any pretreatment which may introduce errors. Alternately carbon electrodes have been used because they have very low intensity spectra and are themselves chemically inert and able to withstand very high temperatures. A typical arc circuit is shown in Figure 5.2. In practice the sample is usually put on the positive electrode (anode). For convenience this is usually the lower electrode. Recently the electrical discharge excitation source has been displaced to a large extent by the use of plasmas.

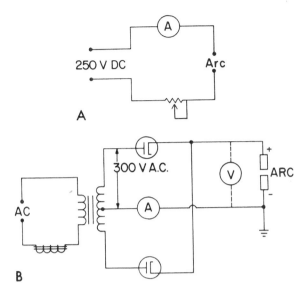

Figure 5.2 Simple DC arc excitation source; (A) DC generator source; (B) rectified AC source.

B. RELATIONSHIP TO FLAME PHOTOMETRY

Emission spectrography was developed after flame photometry. With flames it was clear that the energy source was low and that the procedure was most useful for metals but much less useful analytically for the transition metals or the metaloids.

In emission spectrography an electrical discharge is used to excite the metals of interest. The effective electronic temperature of the electrical discharge is reported to be up to 4,000–6,000 K. Based on the Boltzmann distribution this is clearly a much more efficient atomic excitation system and most metals and metalloids in the Periodic Table can be excited using this technique.

Whereas in flame photometry the most commonly used emission lines are those arising from the transition between the first excited state and the ground state, in emission spectrography these lines are only used when the highest sensitivity is required. At higher concentrations these lines tend to become self absorbed and reverse in intensity. They are said to be reversible lines. This *line reversal* makes quantitative analysis very difficult to interpret. Consequently, nonreversible lines are preferred. These lines arise from transition between two excited states but never the ground state. Since the population of atoms in each of these excited states is always very low, self-absorption is minor and usually a linear relationship is obtained between emission intensity and concentration of the element.

In summary, therefore, emission spectrography has two major advantages over flame photometry. First, it is much more universal in application and, second, it can utilize transitions between upper excited states and measure lines which are not reversible at increased concentrations.

C. EQUIPMENT

1. Optics

A schematic diagram of an emission spectrograph is shown in Figure 5.3. The major components are the electrical discharge for producing the radiation, a monochromator composed of dispersion element and an entrance and exit slit, and a detector and readout system. This simple system is single beam and therefore does not correct for background automatically. As envisioned here, the system would only detect one element at a time, and this would be a severe limitation. In practice, it is highly desirable to detect and measure a number of elements simultaneously. Early systems used a quartz prism.

a. Monochromator

The function of the monochromator, as we know from our earlier discussion (see Chapter 3) is to separate the various lines of a sample's emission spectrum. Both

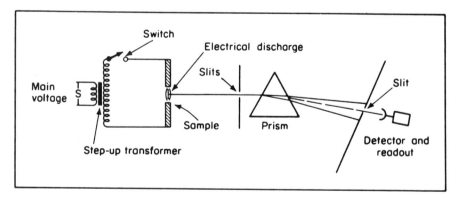

Figure 5.3 Schematic diagram of an emission spectrograph. The lenses are not included in the diagram.

prism and grating dispersion elements are used in emission spectrosgraphy. These are discussed below.

Prism Dispersion Elements   For a prism to be suitable for an emission spectrograph, it must be transparent to UV radiation. For this reason, prisms are usually made of quartz or fused silica. The birefringent property of quartz prisms causes ordinary light to be split into two beams of light, which are polarized perpendicularly to each other.

What appears to be a single beam of light may consist of numerous separate beams, each vibrating in a different plane relative to the direction of propagation. For example, a light beam contains photons, with components oscillating horizontally and vertically (Fig. 5.4).

The act of separating a light beam into two beams vibrating at right angles to each other is called *polarization*. Quartz or fused silica polarize ultraviolet and visible light. The refractive indices of the two polarized beams of light are slightly

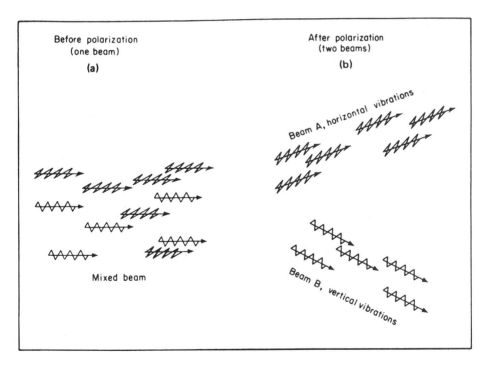

Figure 5.4   Light (a) prior to polarization and (b) after polarization.

different, and the two beams emerge separately from a prism. The result of polarization is that light of one wavelength, such as an emission line, emerges from a prism as two resolved lines. The lines are resolved not because they are of different wavelengths, but because they are polarized. This greatly complicates the interpretation of the spectrum and makes qualitative analysis based on the identification of emission lines very difficult. At the same time, beam splitting into two equal beams causes the loss of half the intensity of each beam, although the total intensity stays constant. This complicates both qualitative and quantitative analysis. The problem can be overcome by using two half-prisms, one of which polarizes light in a right-hand fashion and the other in a left-hand fashion. The first half-prism splits the light into two beams; the second recombines them (Fig. 5.5).

An alternative arrangement is shown in Figure 5.6. This is the Littrow prism, which is essentially a half-prism with a mirrored vertical face. Light enters the sloping face of the half-prism in the conventional manner and is dispersed according to wavelength. In this process the radiation is polarized, resulting in two lines

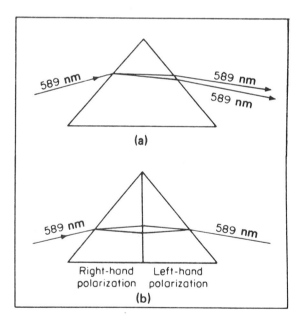

Figure 5.5 (a) An emission line split inot two polarized beams by a quartz prism. (b) The splitting and recombination of an ordinary light beam resolved by two half-prisms.

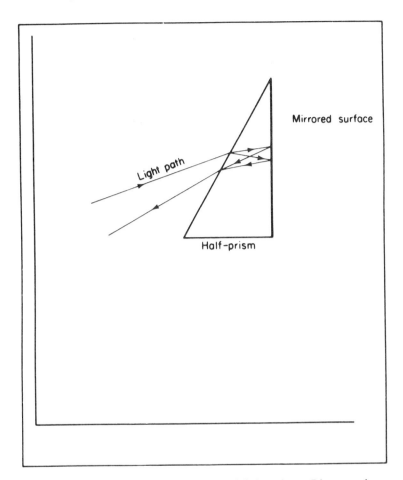

Figure 5.6   Recombination of polarized light using a Littrow prism.

separated very slightly. Each line then reaches the vertical mirrored surface and is reflected back into the half-prism. After reflection, the direction of polarization is reversed. Consequently, the two polarized beams of light are recombined but further dispersed from lines of other wavelengths. The net result is that the radiation emerging from the Littrow prism is resolved according to wavelength, but the splitting caused by polarization is recombined by reflected transmission through the half-prism. With modern instruments ultrapure silica (fluorsil) can be used, which does not polarize the radiation. Consequently, the Littrow prism or the combined half-prisms can be avoided in favor of a simple prism.

Grating Dispersion Elements   Later models of the emission spectrograph used gratings dispersion elements. These had the advantage that the resolution was independent of wavelength, and there were no problems with polarization. In addition there was no loss of radiation by absorption by the grating material—a significant problem whem prisms were used.

A further advantage was that gratings could be curved. These curved gratings can accept a wide-angle beam of radiation and focus it to a point. The focal points of all wavelengths lie on a Rowland circle (see page 250). Photomultipliers can be located at the putative position on the Rowland circle corresponding to the wavelengths monitored for each element. A number of photomultipliers can be used, and a number of elements can be monitored simultaneously. The exit slit must lie on the Rowland circle and may simply be a narrow slit cut in a metal bar permitting radiation to pass through and fall on the detector. An alternate system is to use a very thin mirror, which is located at the putative position, reflecting that radiation into the photomultiplier, which is in the higher or lower plane. This permits clusters of photomultipliers to be located at wavelengths which are very close to each other. The mirror acts as the slit since radiation which does not fall on the mirror is not reflected and does not fall on the photomultiplier.

In practice the position of the photomultipliers must be checked on a daily basis. This can be done by running a standard containing the elements of interest and slowly moving the photomultiplier over the approximate required position and recording the signal. When the signal goes through a maximum, that is the optimum position for the photomultiplier. This can be computer controlled, in which case the scanning and locating place much more rapidly and generally more reliably.

Echelle Monochromators   In recent years echelle monochromators have become increasingly used in emission spectroscopy. Although this technique has been known for many years, it was mostly of academic interest only. However, it has now been incorporated into some commercial plasma emission spectrometers.

The basic premise is to use two dispersive elements in tandem, a prism and a grating. These are placed so that the prism disperses the light in one plane. The dispersed light falls upon the grating, which disperses in a plane at right angles to the plane of dispersion of the prism (Fig. 5.7). The net result is that the final dispersion into wavelengths takes place over a two-dimensional plane rather than along a single linear array as practiced in the Rowland circle optics. The radiation intensity is measured using numerous photomultipliers, which are located on a grid.

It will be appreciated, of course, that the optics are much more complicated than illustrated in the schematic diagram of Figure 5.7, because all focusing lenses and mirrors have been removed for simplicity of illustration. This will be discussed more in the section on plasma spectroscopes.

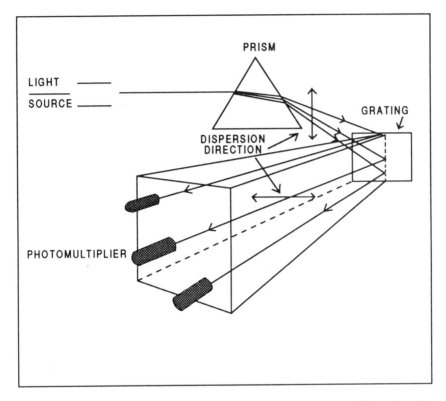

Figure 5.7   Schematic diagram of an echelle monochromator. The prism disperses in the vertical plane; the grating disperses in the horizontal plane.

2. Emission Sources

In general, two types of electrical discharges have been used: the DC arc and the AC spark.

a. Arc Discharge

The DC arc utilizes high current (5–30 amps) and low voltage (10–25 volts). The electrical discharge is struck between two electrodes, one of which contains the sample of interest. Normally the discharge is in air. The temperature of the system ranges between 4,000 and 6,000 K. Metals can be used directly as one of the electrodes, in which case the discharge is rich in the elements in the metal. The emission spectra of the metals in alloys can be obtained directly. This eliminates preparation of the sample. There is a significant time savings. Further errors introduced during treatment are avoided. These include contamination or loss of the sample during transfer.

For other compounds usually the matrix of the compound may vary so greatly that the emission intensity at a constant concentration may vary orders of magnitude. For example, if in the determination of iron in wood the wood was used as an electrode, it is possible to strike an arc but the wood itself would be so volatile that it would burn in the air, obscuring the emission intensity of the iron present. In addition, the iron would be excluded from the discharge by the highly volatile components of the wood, and its intensity would be low. In contrast, if stone or silica were used directly as an electrode and if it were possible to strike an arc between this electrode and a counterelectrode, then it is likely that the iron would be preferentially vaporized and become excited, resulting in more intense iron lines even if the iron concentration were the same as in the wood sample. This problem arises because the matrices of the two samples (wood or stone) are different. It can be overcome by preashing the sample (see page 232).

Arc Stability One problem with the arc is that it tends to wander across the anode. The arc itself tends to strike from the upper electrode (cathode) to the lower electrode, usually impinging on an elevated point on the surface. This point becomes very hot and vaporizes into the electrical discharge, taking with it the sample. After a while the "point" becomes a "pit" and the discharge ceases from that position but wanders to another "point." The same series of events then continues in quick succession. Consequently, the electrical discharge wanders over the surface. The difficulty as far as quantitative analysis is concerned is that the metal of interest may vaporize very rapidly in the first instance and then become depleted with time, giving a lower signal. Then when the arc wanders to another point, the metal emission intensity again increases and then fades off to a lower intensity. This process is continued hundreds, perhaps thousands of times during a single analysis, resulting in a very variable emission intensity on a micro scale in both time and space. The measured intensity is then a function of the length of time that emission is observed. Further, on a macro scale this is not a linear relationship, as will be discussed later.

The DC arc is highly sensitive, and therefore many attempts have been made to improve its reducibility and improve its usefulness for quantitative analysis. Myers and Brunstetter used a rotating magnet near the arc, and Stallwood used a jet of air or inert gas to surround the arc and thus stabilize it. A DC arc using carbon electrodes produce CN bands when operated in the air because the carbon interacts with nitrogen. However using a Stallwood jet, this can be significantly reduced.

b. Spark Discharge

The alternating current spark discharge overcomes a number of the problems experienced with the arc discharge. In this case an AC current is used also. It is built up using an induction coil, a capacitance, and a rotating connection, which provides a surge of electrical energy to the spark gap (Fig. 5.8). The rotating connection provides a short spark of duration 10–100 microseconds with approximately 1,000–2,000 pulses taking place during that time. This greatly averages the conditions of the discharge and so increases considerably the quantitative usefulness of the emitted radiation. A step-up transformer is used to provide a voltage of 10–50 kilovolts as desired.

The AC spark is most useful for quantitative analysis but is less sensitive and therefore has a more limited range than the DC arc for qualitative work.

As a compromise, an alternating current arc can be used by using AC current instead of DC for the typical DC arc set up.

The normal commercial emission spectrograph allows the operator to use whatever conditions he wants for operating of the discharge. This can go from completely DC arc to completely AC spark. In practice the operator selects conditions which are as arclike as necessary to get the sensitivity he requires and as sparklike as possible to get the accuracy and reproducibility required. The selection of conditions is a matter of skill and experience. In recent years this task has been taken on by the instrument manufacturers, who have developed computer programs for numerous types of samples and the metals of importance in them. For

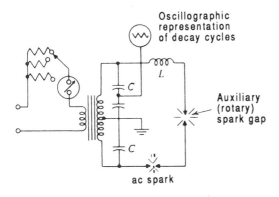

Figure 5.8   Circuit diagram for an AC spark source.

example, in stainless steel analysis (18:8), the user manual will include discharge conditions (arc/spark), wavelength, and discharge times to monitor the nickel and chromium. Another program will describe conditions to determine the major impurities in this alloy. Literally hundreds of such programs have been written and provided by the instrument manufacturers.

3. Electrodes

Most samples are converted to the solid state before being presented to the excitation discharge. The two electrodes used are usually designated the *sample electrode*, normally the lower electrode, and the *counterelectrode*, normally the upper electrode. The sample to be analyzed is put into the sample electrode.

a. Metal Samples as Electrodes

It is frequently possible to analyze metal samples directly. This is because they conduct electricity easily and when used as the sample electrode, volatilize and generate excited atoms and ions. A distinct advantage of the metal sample is that no sample preparation is required other than shaping the sample itself to fit into the electrode.

A common shape is a simple point which is lined up opposite to a pointed counterelectrode. The discharge occurs initially between the two points. This is illustrated in Figure 5.9. For metal sheets the sample may not be round but may actually be cut out of the metal and installed as a flat metal point. Analysis is carried using the shaped metal cutting as the sample electrode.

b. Nonmetallic Samples

Nonmetallic samples are usually pretreated by wet ashing or dry ashing to elim-inate any organic material present. The sample eventually becomes a powder, usually composed of a mixture of metal oxides, often designated M_2O_3. This powder is mixed with a large excess of graphite or silica or some other suitable matrix and the mixture put into the sample electrode. The sample electrodes are designed to hold a small quantity of this mixture in a cuplike shape (Fig. 5.10). The graphite used is usually of high quality and purity so that it does not give emission signals from impurities which would interfere with the analyte signal.

There are a number of shapes suitable for analytical purposes. Usually the spectrographer favors one of these shapes for his own sample types. It is important that the shape used be consistent both for the samples and for the standards used for preparing the calibration curves. Changing shapes can change the volatility of the sample and therefore its response.

c. Electrodes for Liquid Samples

Liquid samples are sometimes quite difficult to handle. If the liquid is organic, it is quite possible that it will burn and the flame will interfere with the emission

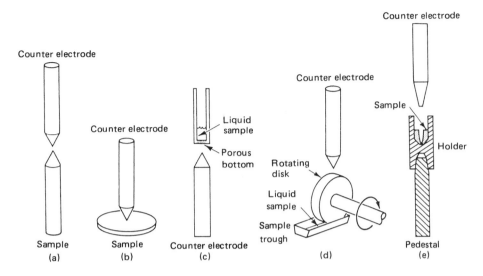

Figure 5.9   Electrode configurations: (a) point-to-point, (b) point-to-plane, (c) porous cup, (d) rotating disk, and (e) carrier distillation.

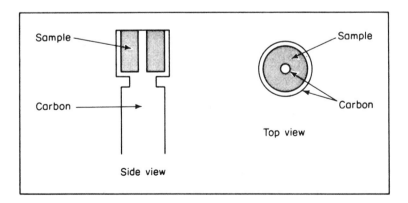

Figure 5.10   Sample holder for powders.

spectrum from the discharge. This interference can take the form of a masking of the emission or the generation of high background by the combustion taking place.

Two designs have been developed over the years for handling liquid samples: the rotating disc and the use of a hollow electrode.

The rotating disc as shown in Figure 5.11 utilizes a carbon disc which dips into a small trough containing the liquid sample. As the disc rotates, the wet surface of the disc moves across the point of the counter electrode and discharge takes place. One problem with this system is that the surface of the rotating disc changes with the number of revolutions, becoming rougher each revolution time. This results in more sample being absorbed onto the surface during rotation and therefore a change in signal intensity against time. A typical response of emission intensity versus time is shown in Figure 5.12. In practice the first few revolutions are ignored and only after the signal has become essentially independent of the number of rotations is the emission intensity measured. The same process is used with the standards used to prepare the calibration curves.

A second type of electrode is the hollow liquid sample electrode as shown in Figure 5.11. The sample is loaded into a hollow electrode, which in this case is the upper electrode. The counterelectrode is the lower electrode. Discharge takes place between the point of the lower electrode and the base of the upper sample electrode. As soon as discharge takes place, the bottom surface of the sample electrode is punctured in a number of small holes and the liquid sample pours through into the electrical discharge. As with the rotating disc, organic solvents can burn causing an interference with the measurement of the intensity of emis-

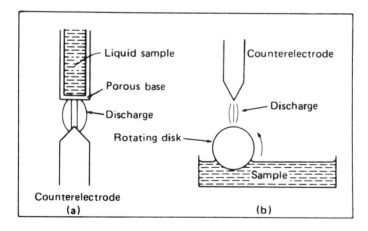

Figure 5.11 Holders for liquid samples: (a) porous cup container and (b) rotating disk electrode.

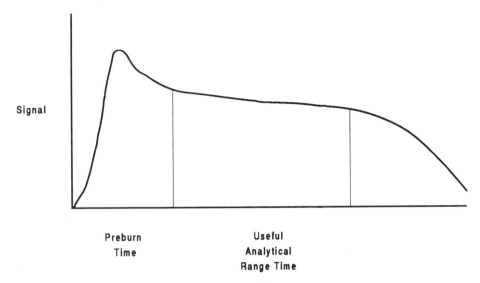

Figure 5.12 Typical response curve for emission intensity versus time of a heavy metal.

sion. A further problem with the hollow sample electrode is that if the solid content of the sample is greater than 1%, as in seawater, urine, or other similar materials, the salt may plate out and plug up the holes in the base of the electrode. The latter then ceases to be porous, the sample ceases to flow, and the signal approaches zero. In general, therefore, high solid content samples must be run on the rotating disc electrode.

4. Detectors

Two detectors are commonly used in emission spectrography, i.e., the photomultiplier and photographic film.

a. The Photomultiplier Detector—Smoothing Condensers

The function of the photomultiplier has already been discussed (see page 75). Unlike its use in obtaining a UV spectrum, it is common for the signal from the photomultiplier to be accumulated at the same wavelength over a period of time rather than read out continuously as the wavelength changes. This may be achieved by keeping the wavelength constant and feeding the signal from the detector into a condenser. As the signal continues to reach the photomultiplier, the output from it increases the charge on the condenser. If the capacity of the condenser is kept constant, then the charge is a direct measure of the total amount of light falling on the detector during the exposure period. This has two important effects. First, it averages out the noise over the entire time period, thus giving a much smoother signal. For this reason we call the condenser a smoothing condenser. Second, the total signal is increased linearly with time so that increased sensitivity is achieved.

Signal-to-Noise Ratio   The detection limit depends on the signal-to-noise ratio, S/N (see page 47). When a smoothing condenser is used, the noise is decreased (but not eliminated) and the signal is increased. However, there is a practical limit to how much S/N ratio can be improved by this means. The improvement in signal-to-noise ratio depends on the square root of the number of observations or time the signal is taken. In order to double the signal-to-noise ratio, we must increase the time exposure by four. This presents us with two problems. First, there must be sufficient sample to give us an extended steady signal, and second, after we get up to a certain exposure, e.g., 4 minutes extended to 16 minutes, to again double the signal-to-noise ratio becomes a major feat. Further extension to 256 minutes would be a prohibitive challenge, and the benefits are not major.

Photomultiplier Response   Various types of photomultipliers respond best over different wavelength regions (see page 75). It is therefore important to select the photomultiplier responding best over the region to be monitored. Typical response curves are shown in Figure 2.13.

b. Film Detectors

For many years film detectors were used for qualitative and quantitative analysis. The use of film for quantitative analysis depended on the darkness of the line compared to the background. As the light intensity increased, the line became darker, and this was measured using a densitometer. However, nowadays film is not commonly used for quantitative analysis. The photomultiplier is more accurate and much faster to read. However film is still the best detector to use for *qualitative analysis*.

D. ANALYTICAL APPLICATIONS

1. Qualitative Analysis

When a sample is brought in with unknown composition, the emission spectrum indicates what metals are present. Interpretation is best done by taking the emission spectrum and recording it on a film which responds throughout the electronic spectral region. Emission is therefore recorded from all elements present in the sample. This greatly eliminates the hazard of unsuspecting metals being present, which would not otherwise be detected using a "guess what's present and test for it" technique. It was then the function of the emission spectrographer to interpret the spectrum and identify the metal or metals that generated this spectrum. This of course is a very difficult procedure and requires a high degree of skill on the part of the operator.

First, it is impossible for the operator to remember all the spectral lines emitted by each particular element. In practice the operator focuses his attention on reading the RU lines. These are described below.

a. Raies Ultima (RU) Lines

If a concentrated sample of a particular element is introduced into a discharge, it will emit many lines. If the concentration of the element is decreased to one-tenth, the weaker lines will not be detectable, but the stronger lines will persist. If it is further diluted to one-tenth of that concentration, more lines disappear. By successive dilutions, the emission spectrum can be decreased until only a very few lines persist. These are termed the Raies Ultima, or RU lines. They are the lines generated by transitions between upper states and the ground state. They are the lines used in atomic absorption spectroscopy for measuring absorption. They are also the last lines to disappear when the concentration of a metal is successively decreased. *Consequently these lines are always present in any sample if that element is present in detectable quantities.* If these lines are absent, then that element cannot be present at detectable concentrations in the sample. Confirmation of the presence of an element then depends on the presence of the RU lines. The operator can therefore ignore all of the other lines in the emission spectrum and concentrate only on the RU lines of the elements present.

Even under these simplified conditions it is often quite difficult to distinguish between the RU lines of one element and other emission lines from a different element which may fall within very short wavelengths of the RU line. Simple examination will not distinguish between such lines. Some other method is therefore necessary to assure identification.

This is done by using a standard film, which contains the emission spectrum of iron and underneath that the RU lines of the various elements of the Periodic Table. From this the operator identifies the wavelengths of the lines emitted by his sample exactly. Then he compares these with the RU lines of the elements he suspects are present. The RU lines are labeled on the standard film.

The practical procedure followed is as follows: (1) The operator takes a photograph of the emission spectrum of the sample to be examined. (2) Then the emission spectrum of iron is recorded on the same plate, but immediately below (1). (3) The film now contains the iron spectrum and the spectrum of the unknown sample. This film is now lined up with the standard film, which contains an iron spectrum and the RU lines of the elements in the Periodic Table. In order to line up the wavelengths, the iron spectra on the standard film and on the unknown film are lined up with each other. This ensures that the two films are now lined up exactly by wavelength. The RU lines may then be checked in the unknown sample and the presence and absence of the various elements can be confirmed.

A more modern approach is to line the photomultiplier up on a particular line known to be present for a particular element (e.g., an RU line). If radiation falls on this photomultiplier, it is assumed that that element is present. In order to do this satisfactorily for a complete qualitative analysis, it would be necessary to have photomultipliers for all the elements in the Periodic Table. This is a practical impossibility. An alternative is to select 20–30 elements and say that these are either present or absent. This process, however, ignores elements other than those monitored. Film is therefore a prime detector for qualitative analysis if all elements are to be included in the search. There is no replacement for a skilled operator.

Another approach is to use only one photomultiplier but program the mono-chromator to successively change the wavelength of radiation falling on it from one RU line to another of a different element. This process of *sequential analysis* is now commonly used in plasma emission spectroscopy.

2. Quantitative Elemental Analysis

In order to set up a quantitative analytical procedure, several factors have to be taken into account before reliable data can be achieved.

First is the choice of the emission spectrum line to be measured. This depends on several factors. These include the expected concentration of the analyte. It is important that the intensity of the line varies significantly over the analytical range in which the analyte concentration falls.

If the concentration is high, the emission line should be weak. The combination of a high concentration and a weak line gives a signal which should be in an acceptable intensity range. If the concentration is low, a weak emission line cannot be used because it may not be detectable. A strong line is therefore preferred. If the concentration approaches the detection limit, an RU should be used since this will be most easily measured.

a. Ashing the Sample

To overcome the problem of different matrices, various kinds of samples are reduced to a common matrix. A procedure often used is first to eliminate volatile organic materials either by dry ashing in an oven at some prescribed high tempe-rature or wet ashing with aqua regia or nitric acid followed by perchlorate to remove the organics. Each method renders the metals to the oxide, often desig-nated MO in the resultant ash. The ash is then mixed with graphite or some other suitable matrix which will be common both to the standards used for preparing the calibration curve and all samples analyzed using this procedure. In that way the matrix of the standards and all of the samples is the same.

In striking the arc the discharge lights upon a high spot in the sample. The

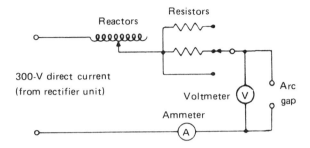

Figure 5.13   Circuit for a DC arc.

intensely high temperature causes vaporization of the matrix and the sample elements into the arc causing an emission of radiation. As described earlier the arc wanders over the surface and an emission intensity which varies wildly on a short time scale is obtained. This can be averaged out over a period of time, resulting in better quantitative results. But the results are always open to question and should be regarded, at best, as accurate to within 5–10% of the true answer.

A circuit for DC arc is shown in Figure 5.13.

A second important variable is that the measured line must not overlap emission lines from the matrix, background, expected impurities, or other components of the sample. Any overlap would lead to a direct error in measuring the line intensity and therefore the reported concentration of the analyte.

A list of emission lines is available in the literature. Lines used in plasma emission are shown in Table 6.3. The best detection limits obtained in emission spectroscopy are shown in Table 5.1.

Other important factors that must be considered when setting up quantitative analysis procedures are as follows.

b. Response Curves

If we put a typical metal into a sample holder and record the emission intensity of a line continuously, we find that the intensity of the emission signal varies with time. This response-versus-time relationship is called a *response curve* (see Fig. 5.12).

The initial part of the response curve is low, increasing to a maximum. This reflects the fact that on striking the discharge first there are no metal atoms emitting in the discharge, but population rapidly increases over a short period of time (several seconds) until a maximum is reached. As the metal from the surface of the sample is depleted into the discharge, more metal must diffuse from the interior of the sample to provide a steady supply of atoms which become excited

Table 5.1 Comparison of Some Experimentally Determined
Emission-Spectroscopic Detection Limits (109/mL)

Element	DC arc[a]	Spark[b]	Element	DC arc[a]	Spark[b]
Ag	0.0006	0.2	Nb	5	0.10
Al	0.05	0.05	Ni	0.02	0.05
As	0.1	5	P	0.15	4
Au	0.05	0.1	Pb	0.005	0.1
B	0.07	0.5	Pd	0.02	0.02
Ba	0.005	0.02	Pt	0.04	0.4
Be	0.0006	0.0002	Rh	0.02	0.05
Bi	0.03	0.1	Sb	0.07	2
Ca	0.01	0.05	Sc	0.2	0.01
Cd	0.02	1	Se	—	—
Ce	0.02	0.3	Si	0.1	0.20
Co	0.01	0.05	Sn	0.05	0.30
Cr	0.01	0.0 5	Sr	0.00003	0.002
Cu	0.0003	—	Ta	30	0.3
Fe	0.01	0.5	Te	60	4
Ga	0.02	0.02	Th	0.02	0.5
Ge	0.02	—	Ti	0.0001	0.01
Hf	1	0.25	Tl	0.07	0.8
Hg	0.07	1	U	—	2
In	0.03	0.3	V	0.02	0.02
La	0.03	0.02	W	0.3	0.4
Mg	0.007	0.05	Yb	0.0009	0.005
Mn	0.003	0.01	Zn	0.01	0.5
Mo	0.006	0.03	Zr	0.004	0.01
Na	0.005	0.1			

[a]Data from V. Svoboda and I. Kleinmann, *Anal. Chem., 40:* 1534 (1968).
[b]Data from J. P. Faris, *Proc. 6th Conf. Anal. Chem. Nucl. Reactor Tech.,* TID-76655.
Gatlingburg, Tenn., 1962.

in the discharge. Eventually the sample is burned out and the signal returns to zero. For quantitative analysis it is important that the emission intensity be as independent of time as possible. Therefore the early part of the emission exposure is usually ignored, since this is often quite erratic. Also the latter part of the emission signal is ignored because this too is not constant with time but falls away to zero. In between these two unacceptable periods the emission intensity is fairly constant versus time. In practice, therefore, it is common to ignore the first period, which may be 8–10 seconds, and the final period. The photomultiplier is therefore programmed to ignore the first part of the emission, called the *preburn time*. Signal is accumulated for the next period of time, where the intensity is reasonably constant and then the discharge cuts off. The sample is removed and a second sample is put in its place. For reproducible data, the preburn time on the standards used for the calibration curve and the samples must be identical, as must the exposure time.

For multielement simultaneous analysis it is very important that the response curves of the metals to be determined be very similar. As can be imagined, the response curves for volatile metals reach a maximum much more quickly than for nonvolatile metals. Such a response curve is shown in Figure 5.14. In practice it

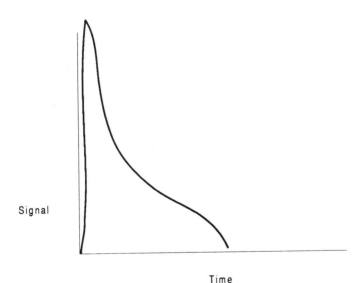

Time

Figure 5.14 Response curve of a volatile metal.

would be impossible to select the preburn time that would be suitable for the quantitative analysis of two such metals simultaneously. By the time the preburn time for the nonvolatile metal had expired, the volatile metal would have been completely burned out and data would not be available. Simultaneous quantitative analysis can only be carried out on metals that have compatible response curves.

As an alternative to this, if it is required that volatile and nonvolatile metals be measured simultaneously, then the preburn time can be ignored and the total signal emitted over a fixed period of time accumulated. This of course will lead to increased error and less accurate results. As long as the operator and the sample provider are aware of this problem and make compensation in interpreting the results, then no harm may be done. This is not a problem in plasma emission spectroscopy where the emission signal is continuous as long as sample is fed into the plasma.

c. Signal Integration

The sensitivity and precision can be improved by integrating the radiation signal over a carefully controlled period. The radiation falling in the detector generates an electrical impulse, which is collected in a storage capacitor. The voltage of the storage capacitor depends on its capacity and the amount of electricity reaching it from the detector. After the predetermined exposure period, which does not include the preburn time, the voltage is read out and related to the radiation intensity falling on the detector. This intensity is in turn related to the sample concentration, which is determined with the help of the calibration curves.

Various techniques have been used by the manufacturers to measure the voltage, such as (1) low capacity capacitor, which are repeatedly discharged and measuring the number of times it is discharged during exposure and (2) high capacity capacitors, the voltage which is measured after exposures.

d. Internal Standards

Packing the sample correctly is most important because this controls the rate at which the sample diffuses into the arc during measurement. This in turn controls the intensity of the emission signal. The calibration curve assumes all samples and the standards have been packed consistently each time and that over the same period of time the same amount of sample is injected into the discharge. This is frequently a source of error. For example, it is not difficult to imagine that the rate at which the sample atoms enter the discharge vary by as much as 10%, resulting in a 10% error in the data acquired. This problem can be overcome to a considerable extent by using an *internal standard*. A typical internal standard is provided when a controlled quantity of a selected element is added to the mixture of sample ash and matrix. A photomultiplier is then set on a line emitted by the element selected as an internal standard. The intensity of that line is a measure of how much internal standard enters the discharge. This in turn is a measure of how much total sample has entered the discharge and is therefore a more accurate way to measure sample exposure rather than simply measuring the length of time over which the discharge has taken place.

Internal standards must be elements which are not present in the original sample, otherwise the signal intensity does not reflect the concentration of the standard metal added to the sample. Gallium is commonly used as an internal standard since it is usually absent from most samples. It is also important that the internal standard not emit at the wavelengths to be monitored during the analysis. It must also have a response curve compatible to the samples being determined.

e. Total Radiation Intensity Measurement

As an alternative to using an internal standard, the *total radiation* from the sample can be used to measure the quantity of sample burned. This can be achieved by a photomultiplier exposed to the entire radiation from the discharge rather than to a single wavelength after it has passed through a monochromator. The premise of this monitoring procedure is that the total radiation from the sample is a measure of the total amount of sample burned. This procedure is certainly more accurate than simply measuring the time of exposure, and it is more convenient in that it does not require the addition of a foreign material to the sample. This saves time and avoids contamination.

3. Interference

In quantitative analysis there are a number of interferences that must be controlled.

a. Spectral Interference

Spectral interference occurs when an element of interest is being monitored at a particular wavelength and another element emits at that same wavelength or a wavelength so close that the photomultiplier is unable to distinguish between the two lines. If this kind of spectral interference occurs, it is often possible to choose a second wavelength to monitor where spectral interference does not occur.

This type of interference can best be verified using film to examine the emission spectrum of the elements in the total sample and the analyte.

b. Background Emission

In the emission discharge there are often molecular species or fragments of molecular species which emit broad-band radiation. Typical of these are the CN bands, which occur when a discharge occurs between carbon electrodes in air. They are caused by the interaction of carbon and nitrogen in the air. Other emission lines will occur from OH bands, CC bands, and CH bands if organic compounds are present. The background radiation may occur at the same wavelengths as the emission line being monitored, in which case a correction must be made. This can be done by measuring the intensity at the wavelengths immediately adjacent to the line being monitored and subtracting the value from the intensity at the monitored wavelength. There are a number of automatic background correctors on the market. One system injects a quartz wedge into the optical path, which has the effect of slightly shifting the wavelength falling on the detector. This may go in and out of the optical path at a fairly high frequency. The photomultiplier therefore sees the background and the background plus the emission line at an alternating rate determined by the oscillation of the wedge. This generates an alternating signal in the photomultiplier, the amplitude of which is the net emission signal and is independent of the background since it is the difference between the background and the background plus emission signal. There are a number of other techniques used for measuring and correcting for background. However, in all cases where quantitative analysis is being performed background must be corrected for in order to get reliable data.

c. Matrix Effects

The matrix is the bulk of the material that makes up the sample presented to the electrode. After mixing, the sample ash (or other form of sample) finds itself surrounded by this matrix. A sample element may be more or less volatile than the matrix. If it is more volatile, it is preferentially atomized and excited and an intense signal is generated for a relatively short period of time. However, if the sample element is less volatile, then the matrix is preferentially volatilized and the sample signal is extended over a longer period of time and at a lower intensity. The matrix and the sample must have compatible volatility in order to give an emission signal with a reasonable response curve. For reproducible results, it is essential that the matrix of the standard and of the sample be identical so that the relative rate of sample volatilization be kept constant (see page 233).

4. Semiquantitative Analysis

By using a rotating disc shaped as in Figure 5.15 and placing it after the exit slit, film can be used to obtain semiquantitative analysis. The rotating disc exposes different parts of the film for different periods of time. If the emission intensity is strong, then even the shortest exposure is sufficient to give a line. However, if the emission intensity is weak, then only the longest exposure generates a line. The intensity, therefore, can be estimated by the length of the line. This technique is useful for semiquantitative analysis but only provides "ballpark" data at best.

5. Analytical Uses for Emission Spectrography

The two major uses for emission spectrography are qualitative and quantitative analysis of the total metal concentration in a sample. There is no indication of speciation of the metal present.

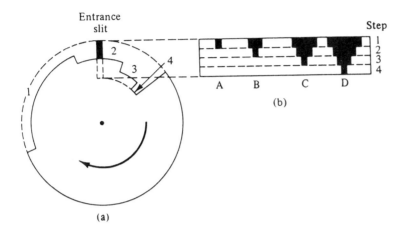

Figure 5.15 (a) Logarithmic step-sector disk rotating in fromt of entrance slit of spectrograph. (b) Schematic illustration of a four-line spectrum obtained with the sector. Line intensity (blackness on emulsion) is indicated by the width of the line.

For quantitative analysis, the fastest and most accurate method is to use a photomultiplier with electronics, which gives an intergrated signal with a reduced signal-to-noise ratio and improved accuracy. The use of film for quantitative analysis has always been very difficult because it included so many steps which must be performed under identical conditions. It was necessary to take the photograph of the standards in the samples under the same conditions and at the same temperature. This is not always easy to do since the conditions in the arc discharge may vary from one sample to the next. Also, it is most important to develop the film under exactly the same conditions of temperature and time. This itself is a work of art rather than science. Consequently, electronic readout systems are much to be preferred.

A unique and most valuable use of emission spectrography is for qualitative analysis. There is no other technique comparable to emission spectrography and plasma emission spectrography for qualitative analysis of metals. This permits simultaneous monitoring of emission spectrum and identifying whatever elements are present quite rapidly. This advantage cannot be duplicated in other techniques such as electrochemistry, atomic absorption, X-ray fluorescence, flame photometry, etc.

The method is also used extensively for quantitative analysis of metals and

Table 5.2 RU Lines of Common Metals

Element	Wavelength of RU lines (nm)	Element	Wavelength of RU lines (nm)	Element	Wavelength of RU lines (nm)
Ag	328.0	Co	345.3	Mo	379.8
	520.9		352.9		386.4
Al	396.1	Cr	425.4	Na	589.5
	394.4		520.6		568.8
As	183.1	Cs	852.1	Ni	341.4
	228.8		894.3		349.2
	237.0	Cu	324.7	P	253 .5
Au	242.1		327.4		255.3
	280.2	Fe	358.1	Pb	405.7
B	249.7		379.1		363.9
	345.1	Ge	265.1	Pt	265.9
Ba	235.5		270.9		299.7
	553.5	Hg	185.0	Se	207.4
Be	234.8		253.6		473.1
	332.1		435.8		474.2
Bi	306.7	K	766.4	Si	288.1
	293.8		769.8		390.5
C	229.6	La	624.9	Sn	317.5
	247.8		593.0		452.4
Ca	442.6	Li	670.7	Sr	460.7
	443.4		610.3		483.2
Cd	228.8	Mg	285.2	Tl	498.1
	340.3		516.7		338.3
Ce	226.3	Mn	403.1	V	318.3
	404.0		403.4		437.9

metalloids. It can easily be used for simultaneous multielement analysis. It is commonplace to determine 30 elements simultaneously using this procedure.

The technique requires a high degree of skill to get reproducible results. Under the best operating conditions, results in the neighborhood of 2% relative standard deviation are achievable, but usually 5% RSD is acceptable. The RU lines used for qualitative analysis are shown in Table 5.2.

6
Plasma Emission

A. INTRODUCTION

Plasma emission spectroscopy is perhaps the newest tool in elemental analysis of the metals. It is a natural extension of flame photometry and emission spectrography (1). It is based on the same physical principles as these techniques; that is, atoms are generated and elevated to an excited state. They then relax to a lower excited state or ground state and in the process emit radiation at characteristic wavelengths. The energy levels involved that are quantized are the same as those discussed in atomic absorption and flame photometry (see pages 11, 193). But in flame photometry the atoms were generated and excited by the energy released by the chemical reaction involved in combustion. In plasma emission the atoms are generated and excited by a plasma torch with much greater energy (higher temperature).

Two principal types of plasma are available: the DC plasma and the inductively coupled plasma (ICP), or RF plasma. A third type is the microwave-induced plasma. This holds the least promise since it is subject to the most variations in quantitative signal. It will therefore not be discussed further in this chapter. The DC plasma was probably first reported by Margoshes and Scribner. The DC arc plasma with an inverted V was first developed by S. C. Valente and W. G. Schrenke in 1970. It was further developed by Elliott, who added a second tungstun cathode to the system and stabilized the arc. The basic design is shown in Figure 6.1.

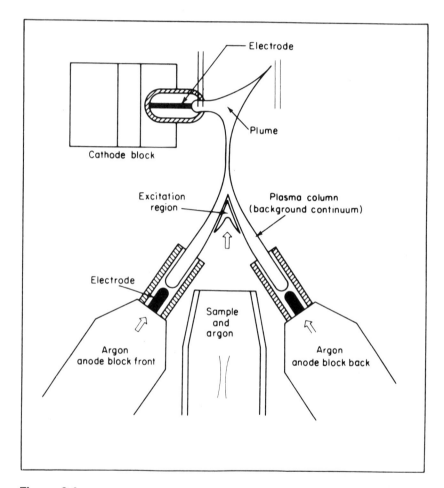

Figure 6.1   Inverted Y configuration of a plasma jet. (With permission from Spectrometrics, Inc., Andover, Mass.)

The RF plasma or inductively coupled plasma (ICP) was first described by A. L. Stolov in 1956. It was studied first by Greenfield in 1964, then studied further by Truitt and Robinson in 1970 and Lichte and Skogerboe in 1972. In 1974, V. A. Fassell and R. V. Kniseley reported the sensitivity of the 61 elements. A common design is shown in Figure 6.2.

As with many analytical procedures, the success of an analytical technique depends very much on its acceptance by the manufacturers. No matter how great the potential use of a new technique, if the manufacturers do not develop a commercial instrument, the technique dies. Fortunately, in this case the manufacturers developed commercialized plasma emission spectrographs which are now commercially available and in common use.

In its simplest concept a stream of argon gas flows through an open tube the end of which is concentric with an RF coil which sets up an oscillating electromagnetic field. A few argon ions are generated using a Tesla tube or some other igniting device. These charged ions are caught up by the oscillating electromagnetic field and try to follow the direction of the field. This causes the ions to oscillate rapidly at the same frequency as the RF. When oscillating the ions are moved rapidly to and fro, in the process striking more argon atoms, which in turn become ionized generating secondary ion pairs of argon ions and electrons. For best performance, the plasma is shaped like a doughnut with a hole in the center of the plasma.

The number of ions rapidly increases until the whole population is highly ionized and a steady state is reached. This constitutes a plasma. The translational energy of the argon ions is very high, and therefore the temperature T in the Boltzmann distribution is very high. But it should be emphasized that these are electronic temperatures. The electronic temperature is measured by the emission intensity. It is assumed that an intensity S would be achieved if the thermal temperature were equal to the electronic temperature T in the Boltzmann distribution, and often no distinction is made. However, it is sometimes important to remember that the temperatures are not really the same.

The electronic temperatures of the plasma are very high. Estimates range from 5,000 to 15,000 K.

However, the sample usually does not go through the hottest part of the plasma but through the hole in the "doughnut." Some estimates are that the temperature is as low as 5,000°C but a consensus holds that the temperature is in the order of 9,000–10,000 K. Even at these "lower" temperatures the emission intensity is very high. Also, by the very nature of the plasma itself, which consist of argon ions and free electrons, it is very easy for free atoms from the sample to become ionized. These ions are also readily excited at the temperatures experienced, and it is not uncommon for ion lines to show the best intensity-versus-concentration relationship.

When atoms of the sample are introduced into the plasma, they collide with the rapidly moving argon ions and become themselves excited. The excited atoms

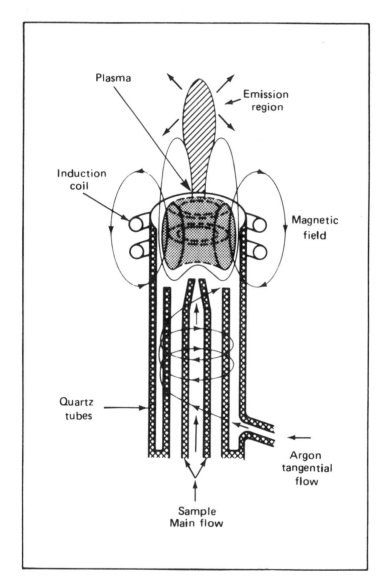

Figure 6.2 Cross section of an inductively coupled plasma torch. Note that the emission region is above the torch.

pass through the plasma, relax to lower energy states, and emit a photon following the equation:

$$\text{Ar}^+ + \text{E}^0 \xrightarrow{\text{plasma}} \text{E}^* \xrightarrow{\text{relaxation}} \text{E}^0 + h\nu \qquad (6.1)$$

atom of   excited

element   element E

The plasma emission torch has a number of advantages over earlier techniques:

1. It is much hotter than the flame or electrical discharge.
2. It is quite capable of exciting all the metals and metalloids in the Periodic Table. Reports have even been made of the detection of atomic oxygen (2) and atomic nitrogen (3) emission.
3. It is much more stable and reproducible than the signal obtained in emission spectrography which employs an electrical discharge, usually across two solid surfaces, one of which contains the sample under investigation.
4. The background is much lower and it suffers from far fewer interference effects.
5. As long as sample is fed into the plasma, the signal is constant; that is, there are no response curves that must be matched for quantitative analysis of multielement analysis.
6. It is capable of simultaneous or sequential multielement analysis. This is a distinct advantage over atomic absorption in spectroscopy, which is best used for single element analysis, although it can be used for simultaneous determination of two or three elements. Instruments designed to determine more than this number simultaneously by AA are most unwieldy.
7. The sensitivity of the plasma is about the same as the sensitivity of flame atomic absorption. It is not as sensitive as carbon atomization flame atomic absorption.

Among the disadvantages of the plasma emission torch is the fact that the background emission may still be significant particularly from the argon present, or any other molecular species introduced with the sample. However, with suitable resolution and background correction, this problem can be minimized.

The most popular system by far is the inductively coupled plasma (ICP), which together with the DCP will be discussed below.

B. EQUIPMENT

1. Inductively Coupled Plasma

A simplistic schematic diagram of ICP equipment is shown in Figure 6.3. It consists of a sample injection system, the argon gases, the induction coil, and the

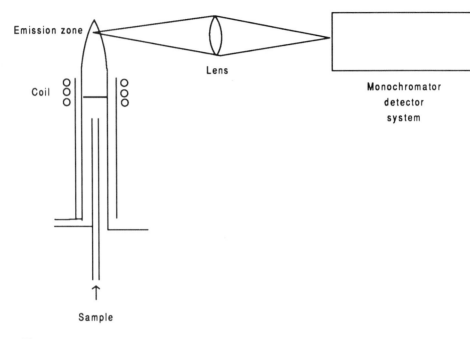

Figure 6.3 Schematic diagram of a typical inductively coupled plasma–optical emission spectroscopy instrument featuring parts of the instrument most important to the user. (From Ref. 18)

emission zone. The radiation from the emission is collected using a lens and focused on various detecting systems. This basic optical system is the same as that used in flame photometry and emission spectrography.

A more detailed description of the ICP system is shown in Figure 6.4. As can be seen, it consists of feed gases, the ICP torch, RF induction coil, sample plume, and a grating polychromator, which separates and directs the pertinent radiation to a series of detectors. These detectors feed the information to a multiplexer and computer system, which generate the quantitative data.

The components of equipment are as follows:

a. RF Plasma Torch

The central component of the ICP is the plasma torch. All other components are standard items in this region of the spectrum.

The early designs of the ICP plasma used three streams of gas flowing through three concentric tubes leading to the RF coil. A cross section of a typical ICP torch is shown in Figure 6.2.

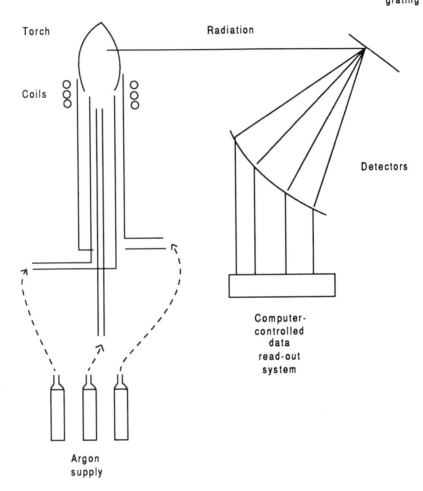

Figure 6.4 Diagram of an ICP emission system detailing the combination of the sequential and simultaneous wavelength detectors in one complete unit. (From Ref. 18)

For many years the only gas used has been argon. This gas is inert, fairly readily available in the industrialized nations, and produces stable plasma with available RF generators. The three streams of gas have three separate important functions. The outer stream flowed at a high rate between 5 to 15 liters per minute and had two functions. First, it primarily sustained the plasma, i.e., argon ions on this stream became the main source of ions in the plasma. Its second and equally important function is to flow fast enough to prevent the plasma itself from contacting the quartz tube. If there is contact, the tube immediately melts and the plasma torch is rendered unusable. In addition to floating the plasma clear of the quartz tube, the fast-flowing argon also conducts away heat and keeps the quartz tube cool. Otherwise the tube would become very hot and melt because of the radiant heat emitted from the torch. The second stream of gas has a variable flow rate. Its function is to control and locate the position of the plasma inside the torch. If the flow rate is too low, the plasma settles too far down the quartz tube and melts it, but if it is too high the plasma is blown outside the RF coils and is extinguished.

Control of the position of the plasma becomes important when the solvents of the sample vary. There is a significant difference in necessary flow rate when the sample changes from aqueous to organic. This can be controlled by controlling the flow rate of the second argon stream.

The third and central gas stream, which flows at a low flow rate between 0.3 and 1.5 meters per minute, is used to convey the gas containing the analyte into the hot core of the plasma. Here the sample is desolvated, decomposed, atomized, and excited. After passage through the plasma, the excited atoms relax in the upper regions of the plasma torch and emit at their characteristic wavelengths. This configuration has been used since the first inception of the process and is still widely used today.

The torch itself was usually made of quartz and comes in various designs. In some designs, the three concentric tubes were fused together by the glass blower to form a single unit. In other designs, the three tubes could be fitted into a plastic end piece so that they were demountable. This permitted the use of alternate materials for the center tube such as boron nitride or aluminum. The outer diameter of the outside tube was conventionally 20 ml, and this size has become standard in the industry.

Considerable attention has been paid recently to the prolific use of argon as the gas producing the plasma. This is because the gas was used in large quantities of up to 18–20 liters a minute and became an expensive item when used over an extended period of time. Many studies have been made to at least replace the argon in the outer stream with nitrogen or air, thus reducing the cost of operation. It was found that this required higher power levels in the RF generator to generate a plasma.

Recent RF generator improvements have permitted further development and the use of N_2 or air. These type of torches are becoming increasingly attractive, especially in underdeveloped nations or in areas of the world where argon is very expensive.

Torches have been developed with low gas consumption. These usually use water cooling or exterior air jets to cool the torch. Some success has been achieved in smoothing the inner torch surfaces. Further, the outer stream of gas is introduced tangentially so that it swirls around the quartz tubing. Increasing this swirl rate has permitted a decrease in gas consumption of as much as 40–50%.

Recently torches have been developed using only nitrogen or air in all three gas streams. Typically these torches require only low kilowatt power but with a frequency of 40.6 MHz. They require increased quantities of sample to generate detection limits similar to the argon plasma torch.

b. Plasma RF Generators

Early generators such as the Lapel RF generator were large (5 feet tall) and stood on the floor. However, with the advent of solid state electronics, smaller desktop systems are now available. Earlier models required a Tesla tube or some other device to initiate the plasma, which then became self-sustaining. This was always an exciting operation generating stress fractures in research equipment, technicians, operators, faculty, and graduate students alike. Automatic ignition is now fairly common.

Several types of circuit design are currently in use. The first is a crystal controlled multistage RF amplifier. The crystal sets a fundamental frequency. That frequency or some multiple of that frequency is tuned with a residence circuit which drives the RF generator. Under standard conditions the plasma then operates in a stable fashion. However, when a sample is introduced into the plasma there is a change in the impedence along the RF transmission line, resulting in an impedence mismatch. This in turn affects the power available and transferred to the torch.

In a second system used in practice, the plasma is kept stable by carefully controlling the flow rates of the various argon streams and the voltage and frequency of the RF generator. However, it is still difficult to control the plasma itself since the introduction of the sample changes the conditions in the plasma.

A third system is the feedback system. In this system a line is monitored and used as a reference indicating the intensity of the plasma. As the line increases in intensity, the power to the RF is decreased bringing the line back to normal. If the line decreases in intensity, the RF is increased and the intensity brought back again. This system, which is a feedback system based on the emission intensity from the plasma, improves the stability of the system. It is particularly useful when the system is warming up, when considerable instability is often experienced. This system has the advantage of stabilizing the system more rapidly than the others (4).

2. DC Arc Plasma

The design of the DC plasma is shown in Figure 6.1. Its function is similar to the system described above. It is a DC plasma jet, which does not use inductive coupling; however, its spectra and analytical features are similar to those of the RF plasma.

3. Optical Systems

Four basic systems have been used extensively in plasma emission spectroscopy. Two of them have been used for many years in the fields of spectrography. These include optics based on the use of a Rowland circle and optics based on the echelle monochromator.

The third system that has been developed extensively for plasma emission uses a single detector, which is kept stationary. It employs a standard metal (usually Hg) as a standard radiation source. Based on the position of a standard line from this element, the monochromator is mechanically rotated to direct radiation of the correct wavelength onto the detector. This is done using a computer-programmed system.

A fourth system keeps the monochromator stationary and mechanically moves the detector to read the intensity of radiation of the required wavelength. These general optical systems are described below.

a. Use of the Rowland Circle

The Rowland Circle is an optical layout which defines the position of the radiation source, a curved grating, and the relevant detectors or exit slits. A schematic diagram of such a system is shown in Figure 6.5.

In this system the radiation source (plasma torch) or an image of the source lies on a circle, which also touches the midpoint of the curve grating and the positions of the photomultiplier detectors. The circle is defined by the radius of curvature of the grating in that the diameter of the Rowland circle is equal to the radius of curvature of the grating.

In practice, the plasma or a focused image of the plasma remains stationary. Radiation falls on the full surface of the grating. The grating resolves the radiation as discussed earlier (pages 69, 222) and refocuses it at various positions on the Rowland circle depending on the wavelength. A photomultiplier is located at the pertinent position of wavelengths to be monitored to measure the light intensity emitted by the sample.

It can be seen that a number of photomultipliers can be used simultaneously and that simultaneous multielement analysis is easily achievable. The signal from the photomultiplier may be fed into a counting device or a condenser which may integrate the signal over a predetermined time. This stored signal from each detector may then be separately read out and based on suitable calibration curves

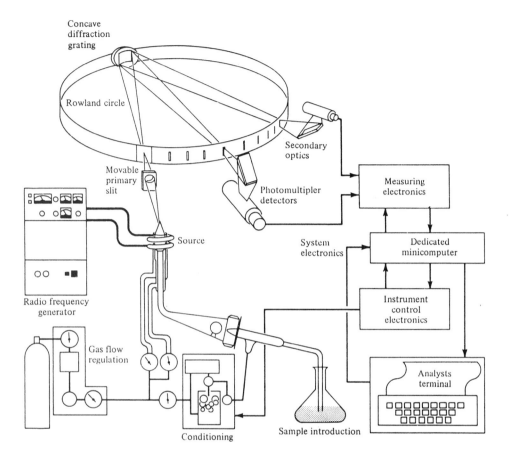

Figure 6.5 Schematic diagram of a nonscanning (direct-reader) spectrometer employing a holographic concave grating in the Rowland circle configuration. (Courtesy of Applied Research Laboratories.)

translated into quantitative data. This of course can all be achieved using a computer system. In practice the photomultiplier may be moved either manually or by computer control to the wavelength of maximum emission from a standard, thus locating it for maximum and reproducible signal measurement.

b. Echelle Monochromator

The echelle monochromator was developed for emission spectrography but not extensively used in commercial instrumentation. It was recognized that resolution using this technique was very high, but physical practical difficulties slowed its general acceptance until the advent of the plasma torch. It was particularly developed for use with a DC plasma. The DC plasma has a somewhat smaller volume in which excitation occurs, and in general this appears to aid in high resolution, a feature which was exploited.

A schematic diagram of the echelle monochromator is shown in Figure 5.7. In this system radiation from the sample falls on a prism, which resolves the radiation by wavelengths in the x direction. The dispersed radiation then falls on a grating physically arranged so that radiation is dispersed in the y direction. The combination generates resolution in two dimensions providing improved resolution compared to the grating alone. Also, the prism eliminates overlapping orders. The latter is an important factor in any system using a grating monochromator. Any group of elements emits at the characteristic wavelengths of each element. Detectors are located on a plane to measure the intensity at each pertinent wavelength. This is done by using a metal plate and drilling holes at the experimentally determined location which corresponds to the dispersed position of the pertinent radiation from the sample, as illustrated in Figure 5.7. Twenty-five elements can comfortably be determined simultaneously with high resolution. More can be determined with special care.

In practice, when a predetermined set of elements are to be determined on a routine basis, a plate is manufactured with holes and photomultipliers in the relative position. The selection of elements is critical and should include all elements to be analyzed routinely and as far as possible any that will be analyzed somewhat regularly. A problem arises if the user later decides to determine another set of metals because it is a case of major surgery to rearrange these holes and photomultipliers into the new locations. Frequently a new plate and detector system must be bought which must be built at the manufacturer's construction site.

As discussed earlier (page 70), the maximum radiation intensity occurs at the blaze angle of a grating and is reduced at all other angles. Instruments are designed so that the blaze angle coincides as closely as possible to the angle of dispersion of the wavelength to be monitored.

In addition, when first order is used, the intensity is also at a maximum, but at higher orders the intensity drops off for a given sample concentration. When several wavelengths are to be monitored as in multielement analysis, clearly all these wavelengths cannot be at the blaze angle. However, by using a number of orders that occur in a different plane from the first order, the blaze angle region can be utilized for wavelengths far removed from the wavelength for which the grating is blazed and some gain in intensity recovered. This is illustrated in Figure 6.6.

Of course this must be taken care of by the manufacturer since the photomultiplier cannot be moved once it is set in place. It is therefore most important that the user designates what elements and what wavelengths he wishes to mon-

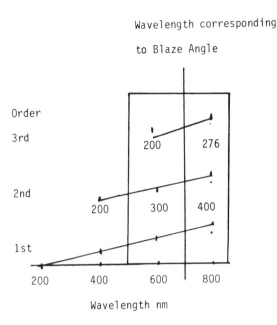

Figure 6.6 Orders stacked above each other in the echelle monochromator system. These are grouped around the blaze angle to give extra light intensity and resolution.

itor. He may not wish to monitor all of them all of the time, so that he should purchase a plate with all those he is likely to use in the foreseeable future.

c. Czerny-Turner

The third optical system uses only one photomultiplier. A schematic diagram is shown in Figure 6.7. This optical system is based on the Czerny-Turner mounting. The system is really a folded Rowland circle with the entrance and exits falling on the circle and the grating falling on an image of the circle. This system has been used in emission spectrography for a number of years quite successfully.

d. Double Monochromator

The demands in improved resolution plasma emisssion has led to the development of a double monochromator or polychomator system. A schematic diagram is shown in Figure 6.8. When a single dispersion element is used, the instrument is computer driven to rotate the monochromator until the desired wavelength falls on the detector. This can be done in a matter of seconds. Using two monochromators

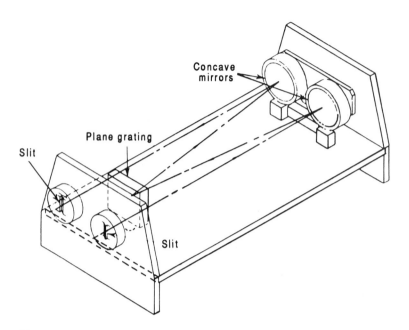

Figure 6.7 The Czerny-Turner mounting.

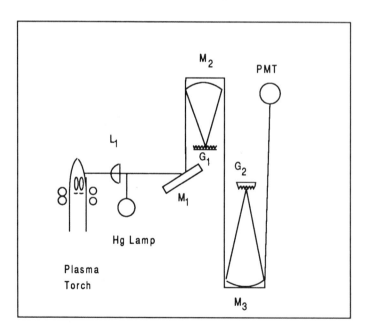

Figure 6.8 Double monochromator with computer-driven gratings is used to minimize stray light effects. Fine tuning for each wavelength is achieved by computer-controlled rotatable refractor plate just before exit slit. Plume observation height is optimized for each element by moving mirror M_1 and lens L_1. G_1, G_2—gratings; PMT—photomultiplier detection; M_1, M_2, M_3—mirrors; L_1—lens.

improves resolution. Each element must be sequentially monitored, rather than being monitored simultaneously as is the case when a Rowland circle and numerous detectors are used. This system basically uses two Czerny-Turner monochromators built in sequence. A potential problem is white light or scatter from the grating. The problem is diminished by using gratings made using interference fringes. Such gratings have low scatter; they also have no blaze angle.

In practice the gratings are used in conjunction with the refractal plate. A mercury emission line is used as the marker. The gratings are driven by a stepper motor at a relatively fast speed until it is at the approximate location to measure the intensity of the emission line. A refractal plate is then driven at a slow speed until the photomultiplier pinpoints the peak by moving slowly across the peak and locating the peak maximum. When locating the desired analytical line, the computer always drives the monochromator in the same direction to avoid problems due to backlash and resultant identification of the wrong emission line as the analytical line. Using this double monochromator, the instrument is capable of determining a number of elements (e.g., 20) in a very short period of time (e.g., 2 minutes). The measurements must be taken sequentially so that the analyses are not quite simultaneous. The lapse time necessary to measure more elements is too great for practical use.

These systems have been produced by commercial manufacturers. Different companies champion different techniques and can point to advantages of each system. Detailed accounts of their operation should be obtained by talking directly to the manufacturer. The "current commercial" systems change yearly and are very dependent on accurate machining, computer handling, and electronic handling.

4. Sample Introduction

Liquids are the most common form of sample to be analyzed by plasma emission. These are usually introduced with a pneumatic spray which injects aerosols into the correct argon stream entering the plasma. With low salt content solutions, the analysis is straightforward. In practice, however, it is always advisable to wash the nebulizer with distilled water after each sample in order to clear out any residue from the previous samples, thus avoiding any memory effect.

If the sample has a high salt content, such as seawater, urine, or other body fluids, it is necessary to use a spray designed to handle high salt content solutions. In practice, an argon saturator nebulizer tip enables washout of the sample and prevents clogging. One interesting method uses a carbon atomizer as a sample introduction system (5).

Aqueous samples are the sample type of choice. Such samples are easy to handle and easily cleaned. Further, the water itself does not generate a very rich spectrum of its own. Organic samples are more difficult to handle since the organic material may give fragments which generate a high background. However, with suitable handling, organics can be run.

Solid samples can be analyzed, but in practice it is difficult to perform quantitative analysis when solid sample introduction systems are used which rely on floating a powder of material into the atomizer.

a. Nebulizer

Several nebulizers are available, including the pneumatic pump, the cross-flow nebulizer, ultrasonic nebulizer, and high solids nebulizers. These have been described earlier. Of these the most popular is the cross-flow nebulizer. A typical schematic diagram is shown in Figure 6.9. A peristaltic pump ensures a steady flow rate of the sample. A cross-flow of gas nebulizes the sample and comes it to the plasma. The design can handle solutions containing high (10%) concentrations of solids, has no memory effect, and the use of a synthetic sapphire tip makes it resistant to corrosion. A recent comparison of the cross-flow nebulizer and the micro Babington nebulizer indicated that the latter was considerably more stable than the former (6).

The thermospray atomizer has been used recently to nebulize liquid samples both for atomic absorption and plasma emission. The advantage of this system is that the sample is reduced to the vapor stage immediately rather than to a system of fine droplets. This eliminates the need for the plasma to evaporate the sample down to the residue and the problem of variable drop size causing variable evaporation efficiency. The system has the potential of significantly improving sensitivity. In preliminary studies it was found that sometimes the detection limits were worse using the thermal spray rather than the normal pneumatic nebulizer (7, 8). This was because of the increase in the noise level and therefore the signal-to-noise ratio and the increase in the standard deviation of the blank sample. However, in some cases the detection limits improved, and this can be explained in terms of improved nebulization efficiency. Further development of this nebulizer is needed.

Electrothermal techniques adapted from atomic absorption have been successfully applied to ICP as a means of introducing a sample into plasma. The technique is particularly useful for organic matrices since it eliminates the effects of organic material. In practice the electrothermal atomizer loosely follows the system used in atomic absorption spectroscopy, drying, ashing, and "atomizing" the sample in sequence. The atoms formed are then introduced into the ICP, where they are further atomized and excited. The method can be used fully with liquid and solid samples and needs only one microgram of material. An excellent account of the current state of the art has been written by Gehart A. Meyer of the Dow Chemical Company.

b. Direct Solids Elemental Analysis

Numerous attempts have been made to introduce solid samples directly into plasma. Some of these have been quite successful, and although efforts have been

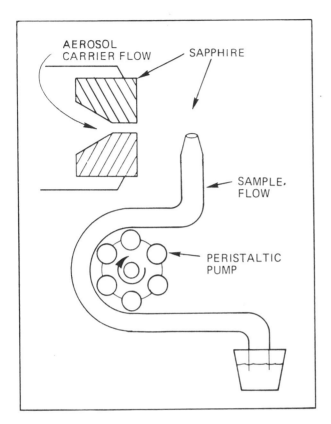

Figure 6.9 Computer-controlled cross-flow nebulizer with sapphire orifice equilibrates the torch quickly, can aspirate solutions with high dissolved solids content, resist corrosion from most samples, and minimize matrix effects.

made to float powdered material into the atomizer this has proved to be difficult. At this time there are no commercial models available with this capability. An account of the state of the art has been presented by Scheeline and Coleman (9).

Another innovation has been the use of fiber optics to convey the image of the plasma to the entrance slit of the spectrometer.

5. Application of Fourier Transform Technique

An extensive study of the determination of metals and nonmetals was carried out by Fateley et al. (10). They looked at the emission spectrum from these elements in the near infrared using a Fourier transformed system. The plasma was run at atmospheric pressure using helium and a microwave-induced plasma (MIP), and it was interfaced between a gas chromatograph and a Fourier transform near infrared (FT-NIR) emission spectrometer. The region scanned was between 15,700 and 7,900 cm^{-1}. The system provided simultaneous multielement analysis for these nonmetals. A combinational technique was developed by Berthoud et al. (11). They determined plutonium in nuclear fuel processing control systems by measuring the intensity of laser-induced fluorescence of the plutonium introduced into an ICP.

Nonmetals have also been determined by coupling a microwave-induced plasma with liquid chromatography by Carnhan et al (15). The separation was carried out using an ion-exchange liquid chromatographics system.

A comparison of the advantages and disadvantages of atomic absorption and plasma emission spectroscopy is shown in Table 6.1. Relative sensitivities of Perkin-Elmer equipment are shown in Table 6.2.

6. Efforts to Reduce the Cost of Operation

The cost of operating the ICP is considerable. Attention has been paid to decreasing this cost. One method was to replace the outer stream of argon with nitrogen or air, which of course is much less expensive. A second approach is to decrease the flow in the outer sheath. Earlier designs used up to 30 liters per minute. Recently designs with total consumption of one liter per minute have been developed, but this involves using a water-cooled torch or directly forced cooling air at the quartz. Argon has been replaced by other gases, but these frequently require higher energy to ionize them in the form of plasma. Recently plasmas have become commercially available with low kilowatt power levels and with a frequency of 40.68 MHz. They can be used with N$_2$ but are usually used with argon.

Small "micro" plasma have become attractive because they are less costly to maintain and seem to be effective as a signal generator.

Helium has also been used in place of argon, particularly for sustaining tiny plasmas operating at 50–100 watts. Helium has a higher excitation potential (18 vs. 14.0 ev) and it is more efficient, therefore, in promoting excitation. However helium is more expensive than argon.

Ideally the best gas for generating the plasmas is krypton, which ionizes at 21 eV; however, it is expensive. Problems have been encountered with nitrogen, which has a very rich spectrum; also it is difficult to maintain the plasma. In practice, the most acceptable gas has been found to be argon.

Table 6.1 Comparison of Absorption and Emission Methods

Basic principle	AA (absorption by ground state atom)	Flame (emission from excited atom)	Emission spectrography (emission from excited atom)	Plasma (emission by excited atom)
Spectral line used	RU line	RU line for best sensitivity / Non-RU line for lower sensitivity	RU line for best sensitivities / Non-RU line for lower sensitiviy	RU lines
Elements	All metals	Group I and II and few others	All metals and metalloids	All metals and metalloids
Analytical Sensitivity	Based on 1% absorption Flame 10^{-6} –10^{-8} 10^{-10} –10^{-14}	Based on S/N ratio 10^{-5} –10^{-6}	Based on S/N ratio 10^{-5} –10^{-8}	Based on S/N ratio 10^{-5} –10^{-8}
Analytical Range	1–2 decades	1–2 decades	1–2 decades	4–5 decades
Interferences	Flame Solvent Chemical Carbon Atomizer Solvent Chemical	Solvent Chemical Spectral	Solvent Chemical Spectral	Solvent Chemical Spectral Matrix
Multielement	Poor	Poor	Yes	Yes
Qualitative	No	Group I and II	Yes	Yes

Table 6.2 Wavelengths Commonly Recommended for Use in Plasma Emission

Element	Wavelength (nm)	Sensitivities	
		Std. Conditions (mg/L)	Opt. Conditions (mg/L)
Ag (silver)	328.068	0.002	0.001
Al (aluminum)	396.152	0.005	0.004
	308.215	0.01	0.01
	167.079	0.05	0.006
As (arsenic)	189.042	0.1	0.05
	193.759	0.1	0.06
Au (gold)	242.795	0.006	0.004
B (boron)	249.773	0.003	0.002
Ba (barium)	455.403	0.0001	0.0001
Be (berylium)	313.042	0.0001	0.00006
Bi (bismuth)	223.061	0.02	0.02
Ca (calcium)	393.366	0.0001	0.00008
Cd (cadmium)	214.438	0.003	0.001
	226.502	0.002	0.001
Ce (cerium)	393.109	0.04	0.03
	413.765	0.01	0.01
Co (cobalt)	238.892	0.003	0.002
	228.616	0.003	0.003
Cr (chromium)	205.552	0.01	0.004
	267.716	0.003	0.002
Cu (copper)	324.754	0.001	0.0009
Fe (iron)	238.204	0.003	0.001
	259.940	0.002	0.002
Ga (gallium)	294.364	0.01	0.01
Ge (germanium)	209.426	0.06	0.05
	265.118	0.02	0.01
Hg (mercury)	194.227	0.05	0.02
In (indium)	230.606	0.04	0.03
Ir (iridium)	224.268	0.02	0.02
K (potassium)	766.490	0.07	0.05
La (lanthanum)	379.478	0.001	0.001
	333.749	0.002	0.001
Li (lithium)	670.781	0.002	0.0009
Mg (magnesium)	279.553	0.0001	0.00008
Mn (manganese)	257.610	0.0004	0.0004
Mo (molybdenum)	202.030	0.02	0.01

Table 6.2   *(Continued)*

Element	Wavelength (nm)	Sensitivities Std. Conditions (mg/L)	Opt. Conditions (mg/L)
Na (sodium)	589.592	0.005	0.004
Nb (niobium)	309.418	0.005	0.003
	269.706	0.005	0.005
Ni (nickel)	231.604	0.007	0.004
P (phosphorus)	177.494	0.1	0.04
	213.618	0.07	0.03
Pb (lead)	220.353	0.04	0.02
Pd (palladium)	340.458	0.002	0.001
Pt (platinum)	214.423	0.03	0.02
	203.646	0.1	0.06
Re (rhenium)	197.313	0.03	0.02
	221.426	0.01	0.006
Rh (rodium)	233.477	0.04	0.02
Ru (ruthenium)	240.272	0.008	0.004
	245.657	0.01	0.008
S (sulfur)	180.731	0.1	0.05
Sb (antimony)	208.833	0.07	0.06
Sc (scandium)	361.384	0.0003	0.0002
Se (selenium)	196.090	0.2	0.06
Si (silicon)	251.611	0.004	0.003
Sn (tin)	189.989	0.1	0.04
Sr (strontium)	407.771	0.00006	0.00005
Ta (tantalum)	226.230	0.03	0.02
Te (tellurium)	214.281	0.06	0.05
Ti (titanium)	334.941	0.0005	0.0005
Tl (thallium)	190.864	0.1	0.08
	276.787	0.06	0.04
U (uranium)	385.958	0.02	0.01
V (vanadium)	292.402	0.003	0.002
W (tungsten)	207.911	0.05	0.02
Y (yttrium)	371.030	0.0003	0.0002
Zn (zinc)	213.856	0.003	0.001
Zr (zirconium)	343.823	0.001	0.001
	339.198	0.0008	0.0008

Source: Courtesy of Perkin-Elmer Corporation.

7. Optical Design of Currently Available Commercial Equipment

a. Perkin-Elmer

Perkin-Elmer offers two instruments. One is high resolution and the other, with lower resolution, is designed to be used for routine analysis only. The latter has a comparatively low resolution holograph grating with about 1800 lines per centimeter and resolution 0.018 nm. The former instrument uses two holograph gratings, one with 1800 and a second with 3400 lines per centimeter and resolution 0.009 nm. The holograph gratings are not blazed. The low resolution grating can be used over the range 160–800 nm, while the high resolution gratings operate over the range 160–400 nm. The gratings are not intended to be used together; rather a selection is made by the microcomputer based on the wavelength to be monitored. The resolution of the Perkin-Elmer Superior model is 0.009 nm, or approximately 0.1Å.

The photomultiplier is kept stationary, and the monochromator system is moved until the desired emission line falls on the detector. The instrument uses one photomultiplier and is therefore sequential. It is able to program quantitative analysis for 70 elements. This gives a wide selection to the user, who is able to make up his own set of elements to be analyzed at any particular time. The monochromators are shown in Figure 6.8.

b. Applied Research Labs (ARL)

ARL manufactures four plasma emission systems. Some are sequential and some are simultaneous in design. All systems use a micro plasma, which reduces the cost of running in terms of argon, which is used at a rate of 8–10 liters per minute. The optical system used is a simple Rowland circle only a small part of which is used in practice.

In the sequential systems, the photomultiplier is kept stationary and the monochromator is moved in sequence to permit the desired wavelengths to fall on the photomultiplier in sequence. This necessitates rigid mechanical control of the system for reproducible results. Drift can become a major problem, but this is reduced considerably by using a very tightly controlled temperature system and constant checking of a standard emission line.

The four marketed models include a fixed position system, which has to be factory made. The customer selects the wavelengths to be monitored, and these allow the simultaneous analysis of about 20 elements within a few seconds. For fully routine analysis this is eminantly satisfactory, but if a different element is to be determined, it cannot be added to the group without major surgery. The cost of this instrument is about $60,000. Their most sophisticated instrument uses an echelle, gives a wide degree of freedom to the operator, and costs about $225,000. The other systems include various features with adjusted market prices.

c. Leeman ICP

This system is a new approach to the optics of ICP. It uses an echelle monochromator but in reverse order, i.e., the grating is used first and disperses radiation in one direction onto a prism which disperses the radiation at right angles. As with all echelle gratings, the light beam is dispersed into a two-dimensional area rather than a circle over which the detection takes place.

The system operates at 40.7 MHz, which gives improved operation of the plasma. In addition, the system has been designed to operate using 7–8 liters per minutes of argon. This is one of the major costs in operating a plasma.

In a major difference from other users, the monochromator system is kept stationary and the photomultiplier detector is moved to different addresses on the echelle graticule for measuring the emission intensity of the desired lines.

Optics The optical system shown in Figure 6.10 is an echelle monochromator

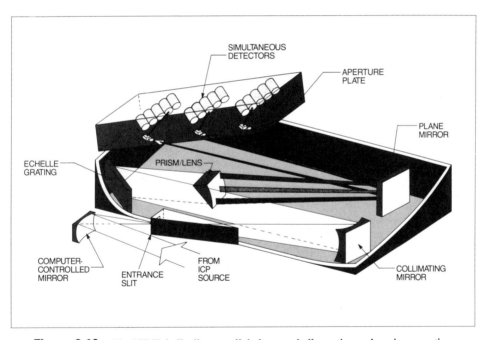

Figure 6.10 The ICP/Echelle disperses light by an echelle grating and a prism, creating a two-dimensional spectrum on the focal plane. Here, a bank of PMTs measures light of specific wavelengths simultaneously. The sequential system uses the same optics, with a PMT that moves between wavelength locations. The PLASMA-SPEC's focal plane lies above the plane of the entry beams, reducing stray light to almost nothing. (Courtesy of Leeman Laboratories, Lowell, Mass.)

wherein radiation falls on an echelle grating where it is dispersed to a prism for dispersion at right angles to that of the grating. The radiation is therefore dispersed in two dimensions as is the normal procedure with echelle gratings.

The dispersed radiation then forms onto a grid of slits as indicated in Figure 6.11. The grid contains literally thousands of "exit slits." Each is rigidly fixed relative to all the others. The photomultiplier is programmed to read the slit relevant to a particular wavelength characteristic of the metal being determined.

Using sequential analysis the photomultiplier moves to one desired slit, intergrates the emission intensity for a predetermined period of time (several seconds) then moves to a second designated slit and repeats this procedure for each element to be determined in the sequence. The selection of wavelengths and exposure time can be manually controlled by the operator who programs it into the controlling computer system.

As an alternate choice to the grid of slits, a fixed apperture plate can be used and a set of photomultipliers located behind each fixed slit. Under these conditions the instrument operates as a simultaneous multielement analyzer.

An advantage of the Leeman system is that the simultaneous or sequential mode can be selected at will.

Temperature control   No temperature control system is used on the Leeman system. Drift is accomodated by lining up a particular slit on a mercury emission line. The whole grid is therefore realigned every 20 minutes in normal operation. This effectively eliminates the need for rigid temperature control, which is otherwise necessary to avoid drift.

Another advantage to the system, of course, is that keeping the monochromators stationary removes one of the major problems in operation, i.e., lack of reproducible control of the monochromator especially with aging.

A list of wavelengths recommended by Leeman and possible interferences are shown in Table 6.3.

C. ANALYTICAL APPLICATIONS

1. Qualitative Analysis

Film is used extensively in emission spectrography, but it is not commonly used in plasma emission. Advances in computerization and tracking have provided alternate procedures for qualitative analysis. One method of qualitative analysis utilized with the ICP is to use a photomultiplier instead of a film.

One procedure is to keep the monochromator stationary and move the photomultiplier slowly around the Rowland circle. A second procedure is to keep the photomultiplier stationary but move the monochromator slowly causing a scanning of the spectral range by the photomultiplier. Either technique has the same effect, but the mechanical problems are somewhat different in each case.

Figure 6.11 Thousands of exit slits are etched into a three-dimensional aperture plate on the focal plane. The rectangle over the plate is the carriage which moved the PMT to the exact location of any emission line, blocking out light from all other lines. (Courtesy of Leeman Laboratories, Lowell, Mass.)

Table 6.3 Prominent Lines Used in Plasma Emission with Detection Limits and Elements that May Cause Spectral Interference

			S/B	<x>	DL 3(x)	Possible Interferences
Ag-1	328.068	I	38.0	10.0	0.007	Fe, Mn, V
Ag-2	338.289	I	23.0	10.0	0.013	Cr, Ti
Ag	243.779	II	2.5	10.0	0.120	Fe, Mn, Ni
Ag-3	241.318	II	1.5	10.0	0.200	
Al-1	309.271	I	13.0	10.0	0.023	Mg, V
Al	309.284	I	13.0	10.0	0.023	Mg, V
Al-2	396.152	I	10.5	10.0	0.028	Ca, Ti, V
Al	237.335	I	10.0	10.0	0.030	Cr, Fe, Mn
Al-3	308.215	I	6.6	10.0	0.045	Mn, V
Al-4	394.401	I	6.3	10.0	0.047	
Al	236.705	I	5.8	10.0	0.051	
Al-5	257.510	I	4.0	10.0	0.075	
Al-6	265.24	I				
As	193.759	I	56.0	100	0.053	Al, Fe, V
As-1	197.262	I	39.0	100	0.076	Al, V
As-2	228.812	I	36.0	100	0.083	Fe, N
As	200.334	I	25.0	100	0.120	Al, Cr, Fe, Mn
As-3	234.984	I	21.0	100	0.142	
Au-1	242.795	I	170.0	100	0.017	Fe, Mn
Au-2	267.595	I	96.0	100	0.031	Cr, Fe, Mg, Mn, V
Au	197.819	I	77.0	100	0.038	Al
Au-3	201.200	I	54.0	100	0.055	
Au	211.068	II	47.0	100	0.063	
Au-4	274.826					
B -1	249.773	I	63.0	10	0.0048	Fe
B -2	249.678	I	53.0	10	0.0057	Fe
B -3	208.959	I	30.0	10.0	0.010	Al, Fe
B	208.893	I	25.0	10	0.012	Al, Fe, Ni
Ba-1	455.403	II	230.0	10	0.0013	Cr, Ni, Ti
Ba-2	493.409	II	130.0	10	0.0023	Fe
Ba-3	233.527	II	75.0	10	0.004	Fe, Ni, V
Ba	230.424	II	73.0	10	0.0041	Cr, Fe, Ni
Ba-4	553.555	II				
Ba	413.066	II	9.1	10	0.032	

			S/B	<x>	DL 3(x)	Possible Interferences
Be-1	313.042	II	110	1.0	0.00027	V, Ti
Be-2	234.861	I	96	1.0	0.00031	Fe, Ti
Be-3	313.107	II	41	1.0	0.00073	Ti
Be-4	249.473	I	8	1.0	0.0038	Fe, Cr, Mg, Mn
Be-5	332.134	I	1.4	1.0	0.021	
Bi-1	223.061	I	87.0	100	0.034	Cu, Ti
Bi-2	306.772	I	40.0	100	0.075	Fe, V
Bi	222.825	I	36.0	100	0.083	Cr, Cu, Fe
Bi-3	206.170	I	35.0	100	0.085	Al, Cr, Cu, Fe, Ti
Bi	195.45	I	14.0	100	0.214	
Bi-4	298.903					
Bi-5	299.334					
C-1	193.091	I	67.0	100	0.044	Al, Mn, Ti
C -2	247.856	I	17.0	100	0.176	Fe, Cr, Ti, V
C -3	199.362	I	3.4	1000	8.823	
Ca-1	393.366	II	89.0	0.5	0.00019	V
Ca-2	396.847	II	30.0	0.5	0.00050	Fe, V, H
Ca-3	315.887	II	5.0	5.0	0.030	Cr, Fe
Ca-4	422.673	I	1.5	0.5	0.010	Fe
Cd-1	214.438	II	120.0	10.0	0.0025	Al, Fe
Cd-2	228.802	I	110.0	10.0	0.0027	Al, Fe, Ni˙
Cd-3	226.502	II	89.0	10.0	0.0034	Fe, Ni
Cd	361.051	I	1.3	10.0	0.230	Fe, Mn, Ni, Ti
Cd-4	326.106	I	0.9	10.0	0.333	
Ce-1	413.765	II	6.2	10.0	0.048	Ca, Fe, Ti
Ce	413.380	II	6.0	10.0	0.050	Ca, Fe, V
Ce-2	418.660	II	5.7	10.0	0.052	Fe, Ti
Ce	393.109	II	5.0	10.0	0.060	Cu (2nd order), Mn, V
Ce-3	404.076	II	4.0	10.0	0.075	
Ce-4	401.239	II	4.0	10.0	0.075	
Co	238.892	II	50.0	10.0	0.0060	Fe, V
Co-1	228.616	II	43.0	10.0	0.0070	Cr, Fe, Ni, Ti
Co	237.862	II	31.0	10.0	0.0097	Al, Fe

(Continued)

Table 6.3 (*Continued*)

			S/B	\<x\>	DL 3(x)	Possible Interferences
Co2	236.379	II	27.0	10.0	0.011	
Co	231.160	II	23.0	10.0	0.013	
Co-3	345.350	II	23.0	10.0	0.013	
Co-4	340.512					
Co-5	350.228					
Co-6	240.725					
Cr-1	205.552	II	49.0	10.0	0.0061	Al, Cu, Fe, Ni
Cr	206.149	II	42.0	10.0	0.0071	Al, Fe, Ti
Cr-2	283.563	II	42.0	10.0	0.0071	Fe, Mg, V
Cr	284.324					
Cr-3	357.869	I	13.0	10.0	0.023	
Cr-4	425.435					
Cr-5	360.533					
Cr-6	427.480					
Cs	455.531	I	0.6+	1000.0	5 0.000	Cr, Fe, Ti, V
Cu-1	324.754	I	56.0	10.0	0.0054	Ca, Cr, Fe, Ti
Cu-2	224.700	II	39.0	10.0	0.0077	Fe,Ni, Ti
Cu	219.958	I	31.0	10.0	0.0097	Al, Fe
Cu-3	327.396	I	31.0	10.0	0.0097	Ca, Fe, Ni, Ti, V
Cu-4	213.598	II	25.0	10.0	0.012	
Cu	223.008	I	23.0	10.0	0.013	
Dy-1	353.171					
Dy	364.540	II	13.0	10.0	0.023	Ca, Fe, Sc, V
Er-1	337.271	II	29.0	10.0	0.010	Cr, Fe, Ni, Ti
Er	349.910	II	17.0	10.0	0.017	Fe, Ti, V
Er-2	390.631	II	14.0	10.0	0.021	
Er	323.058	II	16.0	10.0	0.018	Cu, Fe, Mn, Ti, V
Eu	381.967	II	110.0	10.0	0.0027	Ca, Cr, Fe, Ti, V
Eu	412.970	II	70.0	10.0	0.0043	Ca, Cr, Ti
Eu-1	420.505	II	70.0	10.0	0.0043	Cr, Cu, Fe, Mn, V
Eu	393.048	II	53.0	10.0	0.0057	Ca, Fe, Ti, V
Eu-2	443.560					
Eu-3	368.844					
Fe-1	238.204	II	65.0	10.0	0.0046	Cr, V

			S/B	<x>	DL 3(x)	Possible Interferences
Fe	239.562	II	59.0	10.0	0.0051	Cr, Mn, Ni
Fe-2	259.940	II	48.0	10.0	0.0062	Mn, Ti
Fe	234.349	II	29.0	10.0	0.010	
Fe-3	371.994					
Fe-4	373.713					
Fe-5	248.327					
Ga	294.364	I	64.0	100.0	0.046	Cr, Fe, Mn, Ni, Ti, V
Ga-1	417.206	I	45.0	100.0	0.066	Cr, Fe, Ti
Ga-2	287.424	I	38.0	100.0	0.078	Cr, Fe, Mg, Ti, V
Ga-3	403.298	I	27.0	100.0	0.111	Ca, Cr, Fe, Mn
Gd-1	342.247	II	21.0	10.0	0.014	Cr, Fe, Ni, Ti
Gd	336.223	II	15.0	10.0	0.020	Ca, Cr, Ni, Ti, V
Gd-2	376.840					
Gd	303.284	II	11.0	10.0	0.027	
Gd-3	364.620					
Gd-4	440.185					
Ge	209.426	I	75.0	100.0	0.040	Al, Ca, Cr, Fe, Ni, V
Ge-1	265.118	I	62.0	100.0	0.048	Cr, Fe, Mn, Ti, V
Ge	206.866	I	50.0	100.0	0.060	Al, Cr, Ni, Ti, V
Ge-2	219.870					
Ge	265.158	I	36.0	100.0	0.083	
Ge-3	303.906	I	29.0	100.0	0.103	
Ge-4	326.949					
H	486.133	I	17.0			
Hf-1	277.336	II	190.0	100.0	0.015	Cr, Fe, Mg, Mn, Ni, V
Hf	264.141	II	160.0	100.0	0.018	Cr, Fe, Ti, V
Hf	232.247	II	160.0	100.0	0.018	Fe, Ni
Hf	263.871	II	160.0	100.0	0.018	Cr, Fe, Mn, Ti
Hf-2	273.876					
Hf-3	202.818					
Hg-1	253.652	I	49.0	100.0	0.061	Fe, Mn, Ti

(Continued)

Table 6.3 (*Continued*)

			S/B	<x>	DL 3(x)	Possible Interferences
Hg-2	253.652					
Hg-3	435.835	I	1.7	100.0	2.727	Cr, Cu, Fe, Ni
Hg-4	435.835					
Hg-5	546.075					
Hg-6	365.015					
Hg-7	404.656					
Hg-8	296.728					
Hg-9	194.227					
Ho-1	345.600	II	53.0	10.0	0.0057	Cr, Fe, Ti
Ho	339.898	II	23.0	10.0	0.013	Fe, Cr, Ti
Ho-2	389.102	II	18.0	10.0	0.016	Ca, Cr, Er, Fe, Tm, V
In-1	230.606	II	47.0	100.0	0.063	Fe, Mn, Ni, Ti
In-2	325.609	I	25.0	100.0	0.120	Cr, Fe, Mn, V
In-3	451.131	I	2.1+	10.0	0.140	Ar, Fe, Ti, V
Ir-1	224.268	II	110.0	100.0	0.027	Cr, Cu, Fe, Ni
Ir	212.681	II	100.0	100.0	0.030	Al, Cu, Cr, Ni, V
Ir-2	209.263	I	28.0	100.0	0.107	
Ir-3	215.268	II	44.0	100.0	0.068	Al, Fe, Ni
Ir-4	254.397	I	19.0	100.0	0.157	
K -1	766.490	I	0.5+	1.0	0.060	Ti
K -2	769.896	I	0.4+	1.0	0.080	Cr, Ti
K	404.414	I	1.0+	1000.0	30.000	Ca, Cr, Fe, Ti
La	313.749					
La-1	379.478	II	30.0	10.0	0.010	Ca, Fe, V
La-2	408.672	II	30.0	10.0	0.010	Ca, Cr, Fe
La	412.323	II	29.0	10.0	0.010	
La-3	394.911					
Li-1	670.784	I	1.7+	0.1	0.0018	V, Ti
Li-2	610.362	I	0.8+	1.0	0.038	Ca, Fe
Lu	460.268					
Lu-1	261.542	II	150.0	5.0	0.0010	Al, Ca, Cr, Fe, Mn, Ni,k V
Lu-2	291.139	II	24.0	5.0	0.0062	Cr, Fe, Ti, V

			S/B	\<x\>	DL 3(x)	Possible Interferences
Lu	219.554	II	18.0	5.0	0.0083	Cr, Cu, Fe, V
Mg-1	279.553	II	195.0	1.0	0.00015	Fe, Mn
Mg-2	280.270	II	100.0	1.0	0.00030	Cr, Mn, V
Mg-3	285.213	I	19.0	1.0	0.0016	Cr, Fe, V
Mg-4	383.231	I	0.7	1.0	0.042	
Mg-5	277.983	I	0.6	1.0	0.050	
Mg-6	278.297					
Mn-1	257.610	II	220.0	10.0	0.0014	Al, Cr, Fe, V
Mn-2	259.373	II	190.0	10.0	0.0016	Fe
Mn	260.56 9	II	145.0	10.0	0.0021	Cr, Fe
Mn-3	293.306	II	22.0	10.0	0.013	
Mn	279.482	I	24.0	10.0	0.012	
Mn-4	279.827	I	18.0	10.0	0.016	
Mn-5	403.076	I	6.8	10.0	0.044	
Mo-1	202.030	II	38.0	10.0	0.0079	Al, Fe
Mo	203.844	II	24.0	10.0	0.012	Al, V
Mo	204.598	II	24.0	10.0	0.012	Al
Mo-2	281.615	II	21.0	10.0	0.014	Al, Cr, Fe, Mg, Mn, Ti
Mo	201.511	II	16.0	10.0	0.018	
Mo-3	379.825					
Mo-4	386.411					
Mo-5	313.259					
Na-1	588.995	I	101.0	100.0	0.029	Ti (2nd order)
Na-2	5898.592	I	43.0	100.0	0.069	Fe, Ti, V
Na-3	330.237	I	1.6	100.0	1.875	Cr, Fe, Ti
Na	330.298	I	0.7	100.0	4.285	
Nb-1	309.418	II	83.0	100.0	0.036	Al, Cr, Cu, Fe, Mg, V
Nb-2	316.340	II	75.0	100.0	0.040	Ca, Cr, Fe
Nb	313.079	II	60.0	100.0	0.050	Cr, Ti, V
Nb	269.706	II	43.0	100.0	0.069	Cr, Fe, V
Nb-3	410.092					
Nb-4	202.932					

(*Continued*)

Table 6.3 (*Continued*)

			S/B	<x>	DL 3(x)	Possible Interferences
Nd	401.225	II	59.0	100.0	0.050	Ca, Cr, Ti
Nd-1	430.358	II	40.0	100.0	0.075	Ca, Fe
Nd	406.109	II	31.0	100.0	0.096	Ca, Cr, Fe, Mn
Nd-2	415.608	II	28.0	100.0	0.107	Ca, Fe
Ni-1	221.647	II	29.0	10.0	0.010	Cu, Fe, V
Ni-2	232.003	I	20.0	10.0	0.015	Cr, Fe, Mn
Ni-3	231.604	II	19.0	10.0	0.015	Fe
Ni	216.556	II	17.0	10.0	0.017	Al, Cu, Fe
Ni-4	341.476	I	6.2	10.0	0.048	
Ni-5	352.454	I	6.6	10.0	0.045	
Ni-6	361.939					
O	436.830	I	3.2			
Os-1	225.585	II	83.0	1.0	0.00036	Cr, Fe, Ni
Os-2	228.226	II	48.0	1.0	0.00063	Fe
Os	189.9		25.0	1.0	0.0012	Cr
Os	233.680	II	24.0	1.0	0.0012	Fe, Ni
Os-3	222.798	I	11.0	1.0	0.0027	
P -1	213.618	I	39.0	100.0	0.076	Al, Cr, Cu, Fe, Ti
P -2	214.914	I	39.0	100.0	0.076	Al, Cu
P -3	253.565	I	11.0	100.0	0.272	Cr, Fe, Mn, Ti
P	213.547	I	8.5	100.0	0.352	Al, Cr, Cu, Ni, Ti
P	203.549					
P -4	255.328	I	5.2	100.0	0.576	
P -5	253.401	I	3.0	100.0	1.000	
P -6	255.493					
Pb-1	220.353	II	70.0	100.0	0.042	Al, Cr, Fe
Pb	216.999	I	33.0	100.0	0.090	Al, Cr, Cu, Fe, Ni
Pb-2	283.307	I	21.0	100.0	0.142	Cr, Fe, Mg
Pb-3	280.200					
Pb-4	405.782					
Pb-5	368.347					
Pd-1	340.458	I	68.0	100.0	0.044	Fe, Ti, V
Pd	363.470	I	55.0	100.0	0.054	Ar, Fe, Ni, Ti
Pd-2	342.124	I	30.0	100.0	0.100	
Pd-3	244.791	I	23.0	100.0	0.130	

			S/B	\<x\>	DL 3(x)	Possible Interferences
Pr-1	406.282					
Pr-2	398.972					
Pr	390.844	II	8.1	10.0	0.037	Ca, Cr, Fe, V
Pr	414.311	II	8.0	10.0	0.037	Fe, Ni, Ti, V
Pr	406.281	II	6.3	10.0	0.047	
Pr	398.972					
Pt-1	214.423	II	100.0	100.0	0.030	Al, Fe
Pt	203.646	II	54.0	100.0	0.055	Al, Cu, Fe
Pt	204.937	I	42.0	100.0	0.071	Al, Fe, Ti, V
Pt-2	265.945	I	37.0	100.0	0.081	Cr, Fe, Mg, Mn, V
Pt-3	306.471	I	25.0	100.0	0.120	
Pt-4	299.797					
Pt-5	262.803					
Rb-1	780.023	I	0.2+	1.0	0.150	Ti
Rb-2	794.760	I	0.2+	100.0	15.000	Ar
Rb	420.185	I	1.2+	1000.0	25.000	Fe, Mn, Ni, V
Re	197.313	II	49.0	10.0	0.006	Al, Ti
Re-1	221.426	II	47.0	10.0	0.006	Cu, Fe, Mn
Re-2	227.525	II	44.0	10.0	0.006	Ca, Fe, Ni
Re	189.836	II	8.0	10.0	0.037	Fe
Re-3	346.472					
Rh	233.477	II	67.0	100.0	0.044	Cr, Fe, Ni, Ti, V
Rh-1	249.077	II	52.0	100.0	0.057	Cr, Fe, Mn
Rh	343.489	I	50.0	100.0	0.060	V
Rh-2	369.236	I	35.0	100.0	0.085	
Rh-3	332.309					
Rh-4	343.489	I	50.0	100.0	0.060	V
Ru-1	240.272	II	100.0	100.0	0.030	Cr, Fe, V
Ru	245.657	II	100.0	100.0	0.030	Fe
Ru-2	267.876	II	83.0	100.0	0.036	Cr, Fe, Mn, V
Sb-1	206.833	I	91.0	100.0	0.032	Al, Cr, Fe, Ni, Ti, V
Sb-2	217.581	I	68.0	100.0	0.044	Al, Fe, Ni
Sb	231.147	I	49.0	100.0	0.061	Fe, Ni
Sb-3	252.852	I	28.0	100.0	0.107	Cr, Fe, Mg, Mn, V
Sb-4	259.806					

(Continued)

Table 6.3 (*Continued*)

			S/B	<x>	DL 3(x)	Possible Interferences
Sc-1	361.384	II	200.0	10.0	0.0015	Cr, Cu, Fe, Ti
Sc	357.253	II	150.0	10.0	0.0020	Fe, Ni, V
Sc	363.075	II	140.0	10.0	0.0021	Ca, Cr, Fe, V
Sc-2	402.369					
Se	196.090	I	40.0	100.0	0.075	Al, Fe
Se-1	203.985	I	26.0	100.0	0.115	Al, Cr, Fe, Mn
Se-2	206.279	I	10.0	100.0	0.300	Al, Cr, Fe, Ni, Ti, V
Si-1	251.611	I	250.0	100.0	0.012	Cr, Fe, Mn, V
Si	212.412	I	180.0	100.0	0.016	Al, V
Si-2	288.158	I	110.0	100.0	0.027	Cr, Fe, Mg, V
Si-3	250.690	I	100.0	100.0	0.030	Al, Cr, Fe, V
Si-4	251.921					
Sm-1	359.260	II	69.0	100.0	0.043	Cr, Fe, Ti, V
Sm	442.434	II	55.0	100.0	0.054	Cr, Ca, Ti, V
Sm-2	356.826					
Sm-3	443.432	II	36.0	100.0	0.083	
Sn	189.989	II	120.0	100.0	0.025	
Sn-1	235.484	I	31.0	100.0	0.096	Fe, Ni, Ti, V
Sn	242.949	I	31.0	100.0	0.096	Fe, Mn
Sn-2	283.999	I	27.0	100.0	0.111	Al, Cr, Fe, Mg, Mn, Ti, V
Sn-3	224.605	I	25.0	100.0	0.120	
Sn-4	380.100					
Sn-5	317.502					
Sr-1	407.771	II	72.0	1.0	0.00042	Cr, Fe, Ti
Sr-2	421.552	II	39.0	1.0	0.00077	
Sr	216.596	II	36.0	10.0	0.0083	Al, Fe, Ni
Sr-3	460.733	I	4.4	10.0	0.068	
Ta-1	226.230	II	120.0	100.0	0.025	Al, Fe
Ta-2	240.063	II	105.0	100.0	0.028	Cr, Cu, Fe, V
Ta	268.517	II	100.0	100.0	0.030	Cr, Fe, Mn, V
Tb-1	350.917	II	130.0	100.0	0.023	Cr, Fe, Ti, V
Tb	384.873	II	54.0	100.0	0.055	Ca, Cr, Mg, V

			S/B	\<x\>	DL 3(x)	Possible Interferences
Tb	367.635	II	50.0	100.0	0.060	Ca, Cr, Fe, Mn, Ti,V
Tb-2	370.285					
Tb-3	432.648					
Te	214.281	I	73.0	100.0	0.041	Al, Fe, Ti, V
Te	225.902	I	17.0	100.0	0.041	Fe, Ni, Ti, V
Te-1	214.725	I	14.0	100.0	0.214	Al, Cr, Fe, Ni, Ti, V
Te-2	208.116	I	11.0	100.0	0.272	
Th-1	283.730	II	46.0	100.0	0.065	Cr, Fe, Mg, Ni, V
Th	283.231	II	42.0	100.0	0.071	Cr, Fe, Mg, Ti
Th-2	401.913	II	36.0	100.0	0.083	Ca, Cu, Mn
Th-3	439.111					
Ti-1	334.941	II	79.0	10.0	0.0038	Ca, Cr, Cu, V
Ti-2	336.121	II	57.0	10.0	0.0053	Ca, Cr, Ni, V
Ti	323.452	II	56.0	10.0	0.0054	Cr, Fe, Mn, Ni, V
Ti-3	337.280	II	45.0	10.0	0.0067	Ni, V
Ti-4	364.268					
Ti-5	500.721					
Tl	190.864	II	74.0	100.0	0.040	Al, Ti
Tl-1	276.787	I	25.0	100.0	0.120	Cr, Fe, Mg, Mn, Ti, V
Tl	351.924	I	15.0	100.0	0.200	Cr, Fe, Ni, V
Tl-2	377.572	I	13.0	100.0	0.230	Ca, Fe, Ni, Ti, V
Tl-3	352.943	I	1.7	100.0	1.764	
Tm	313.126	II	58.0	10.0	0.0052	Cr, Ti, V
Tm-1	346.220	II	37.0	10.0	0.0081	Ca, Cr, Fe, Ni, V
Tm-2	370.026	II	14.0	10.0	0.021	
Tm-3	409.418					
U -1	385.958	II	12.0	100.0	0.250	Ca, Cr, Fe
U -2	367.007	II	10.0	100.0	0.300	Fe, Ni, Ti, V
U -3	424.167	II	6.5	100.0	0.461	
U -4	358.488					
V -1	309.311	II	60.0	10.0	0.0050	Al, Cr, Fe, Mg
V -2	310.230	II	47.0	10.0	0.0064	Fe, Ni, Ti

(Continued)

Table 6.3 (*Continued*)

			S/B	<x>	DL 3(x)	Possible Interferences
V -3	292.402	II	40.0	10.0	0.0075	Cr, Fe, Ti
V	290.882	II	34.0	10.0	0.0088	Cr, Fe, Mg, Mo
V -4	437.924					
W -1	207.911	II	100.0	100.0	0.030	Al, Cu, Ni, Ti
W	224.875	II	67.0	100.0	0.044	Cr, Fe
W	218.935	II	65.0	100.0	0.046	Cu, Fe, Ti
W -2	202.998	II	40.0	100.0	0.075	
W -3	276.427					
W -4	400.875					
Y -1	371.030	II	86.0	10.0	0.0035	Ti, V
Y	324.228	II	67.0	10.0	0.0045	Cu, Ni, Ti
Y -2	377.433	II	57.0	10.0	0.0053	Fe, Mn, Ti, V
Y -3	410.238					
Yb	328.937	II	170.0	10.0	0.0018	Cu, Fe, Ti, V
Yb-1	369.419	II	100.0	10.0	0.0030	Ca, Fe, Mn, Ni, Ti, V
Yb	389.138					
Yb-2	398.799					
Zn-1	213.856	I	170.0	10.0	0.0018	Al, Cu, Fe, Ni, Ti, V
Zn-2	202.548	II	75.0	10.0	0.0040	Al, Cr, Cu, Fe, Mg, Ni
Zn-3	206.200	II	51.0	10.0	0.0059	Al, Cr, Fe, Ni, Ti
Zn-4	481.053	I	1.3	10.0	0.230	
Zr	343.823	II	42.0	10.0	0.0071	Ca, Cr, Fe, Mn, Ti
Zr-1	339.198	II	39.0	10.0	0.0077	Cr, Fe, Ti, V
Zr	257.139	II	31.0	10.0	0.0097	Cr, Fe, Hf, Mg, Mn, Ti, V
Zr-2	360.119					
Zr-3	407.270					

Source: Courtesy of J. Leeman.

The location of the photomultiplier identifies the wavelengths at which emission occurs. By programming in the wavelengths of the RU lines of the various elements, they can be identified and reported by the system. This eliminates the photographic plate and its attendant woes. A problem is spectral interferences, which can give false signals for different elements. Interpretation of spectra depends largely on the skill of the operator, although computers have been used in an attempt to eliminate this need for skill.

2. Quantitative Analysis

One of the very attractive features of the plasma is that it has a very wide analytical range, sometimes extending up to five orders of magnitude. This is a distinct advantage over atomic absorption, which has a fairly narrow analytical range rarely greater than two orders of magnitude and usually only one order of magnitude. The wide analytical range allows more concentrated samples to be analyzed and removes the necessity for dilution of the sample. This is always a distinct advantage since any treatment of the sample can lead to operator error and error caused by the introduction of impurities during treatment. The definition of "detection limits" used with plasma emission spectrography is usually twice the standard deviation of the noise level, i.e., when the signal is equal to twice the standard deviation of noise, then it is "detectable." For quantitative work the lower end of the scale is usually five times the detection limits. At low concentrations most elements use the RU line only, but at higher concentrations other elements must be used. Spectral interferences of course are possible, particularly with elements that have rich spectra such as the transition elements. Using computer control and the narrow slits this can be significantly reduced, but it is always a potential source of error.

In plasma emission, the electronic temperature is very high but the plasma is very thin. As a consequence emission efficiency is high, but self-absorption (by ground state atoms) is low. It is therefore acceptable and even desirable to use RU lines for measurement. This ensures maximum sensitivity. However, ion lines have been used extensively because these do not self-absorb. Also, they occur in a region of the plasma where chemical reaction does not occur (decomposition and recombination can occur below and above this region, respectively). As a further consequence of the thin atom population, the plasma emission-concentration relationship is dynamic over several orders of magnitude. Four orders of magnitude is not uncommon. This is a distinct advantage over emission spectrography and flame photometry where the linear range is orders of magnitudes less.

a. Calibration Curves

The use of the calibration curve is based on the premise that the samples and the standards used to prepare the calibration curves are very similar in composition and that no compounds or elements are present which will affect the relationship between emission intensity and concentration of the analyte. This assumption is not always valid, but it is in the evaluation of this assumption that the skill of the operator is called upon. Such changes in relationships can be brought about by the analytical interferences discussed above.

First the operator must select an emitted line with a wavelength remote from other emission lines from the source. In flame photometry the selection is limited by the low energy of the flame and consequently only a few emission lines being available. It is not uncommon to use the RU lines for the determination of the lowest concentrations since these lines are always present in flames and at low concentrations are usually the most intense. In emission spectrography using arc or spark discharge, the temperatures are considerably higher. The RU lines are usually avoided except in cases where the concentration of the analyte is close to the detection limits.

Since the plasma has such a high effective electronic temperature, there is no problem in electronically exciting any of the elements in the Periodic Table. By the same token elements such as sodium and potassium which have RU lines at long wavelengths are also easily ionized. It is therefore common to use ion lines rather than atomic lines to measure these elements.

The echelle monochromator uses both a prism and a grating, but if only one dispersion element is used the grating is invariably used. Gratings have numerous advantages over prisms as has been discussed earlier (page 69). Filters cannot be used because they allow a band pass that is too broad to isolate the desired emission lines. The characteristics of grating should be reviewed if necessary (see pages 69, 222).

3. Analytical Interferences

a. Spectral Interferences (Interelement and Molecular Emission)

The spectra of the elements are very rich, particularly those of the transition elements. Frequently it is necessary for very high resolution to separate lines of different elements from each other. In practice, care is taken to monitor a line with a wavelength that does not coincide with wavelengths emitted by any other element present in the sample. However this is not always possible, in which case another element or several other elements can emit at the wavelength being monitored. This is a spectral interference. Spectral interference is encountered if other elements emit at the same wavelength as the wavelengths being monitored for the analyte.

b. Background Correction

By measuring the intensity of the emission from a number of elements, their contribution to the intensity of radiation at the characteristic wavelengths of other

Table 6.4   Background Correction (Plasma)

	Al (ppm)	Cu (ppm)
Al added	5000	
Cu added		5000
Effective background at Fe line	2	10
Effective background at Ni line	0.2	1

elements can be measured and used to correct the data. This enables automatic correction to be applied to interelement radiation interference with a number of elements. This is particularly important at trace levels and when numerous elements are present in the sample and each may contribute to each other's signal. It should be noted in passing that the ultimate in this approach would be achieved when interfacing radiation from all elements in the Periodic Table is included in the correction. A formidable task, remembering that the intensity of each varies with different matrices.

Some examples of background corrections are given in Table 6.4. Note that if 5000 ppm of aluminum is present, then the signal at the iron wavelength is equivalent to 2 ppm of iron and 0.2 ppm of nickel. Similarly, 5000 ppm of copper will give a signal equivalent to 10 ppm of iron and 1 ppm of nickel at their respected wavelengths.

Assuming that this relationship holds constant, it is possible to set up a matrix for all the elements and calculate the contribution each one makes to each of the others and, by solving the pertinent equations, to produce a corrected signal for each element. This of course can be done with a computer in a short period of time. Fischer Scientific has done this for 50 elements so that quite reliable background correction can be made for radiation interference. It should be remembered, however, that there are more than 50 elements in the Periodic Table, and the other elements will certainly contribute even though they are not corrected for.

If the matrix changes, e.g., if we use an organic solvent instead of aqueous solvent or if we change from sulfate to nitric acid or hydrochloric acid, then there is a change in the contribution of a particular metal at the wavelengths mentioned for other metals and therefore the background correction necessary for accurate answers.

Double Monochromator System to Reduce Background   Stray light is always a significant source of error. Any stray light that falls on the detector is registered as a signal whether it is at the correct wavelength or not. Stray light can come from stray radiation in the system such as that scattered from the surface of a grating or by scattering from high concentrations of analyte in the sample. For example, a solution of 10,000 ppm of Mg will appear to scatter like a solution of 100 ppm if the rejection of the grating is 1000. If two gratings are used, the scatter will be reduced to a level equivalent to 0.1 ppm—a negligible level. It can be greatly reduced by using a double monochromator, as shown in Figure 6.8.

Another spectral interference is background emission from molecular fragments, which results in broadband radiation that can easily overlap many atomic lines. In flame photometry and in arc spark emission spectrography, background emission is frequently very strong. In flame photometry there is usually always emission from OH molecules and other fragments of water decomposition products which emit strongly in flame. These broadband spectra frequently overlap the line emission from the analyte.

Similar problems occur in arc spark spectrography. Here not only do we get emission from material introduced by the sample, but frequently there is a reaction between the electrical discharge and the air in the vacinity of the discharge. The latter results in emission bands from OH ions and NO, NH, and CN fragments. These bands may be quite intense and are often frequently a serious problem in quantitative analysis.

However, in plasma emission there are no electrodes holding the sample, and therefore there is very much reduced background emission from OH, NO, NH, and CN moieties. Consequently the background problem is much reduced in the plasma emission compared to other forms of emission spectroscopy such as arc spark and flame photometry.

In plasma emission two effects greatly reduce interelement and background interferences. One is the very high temperature of this system, which greatly reduces the chemical effect of one element on another. Second, the residence time of an analyte atom in the plasma is about 2 milliseconds. This is a very long period of time to be held at this temperature and results in complete breakdown to the constituent elements of any molecules that may be present. This itself greatly reduces the interelement effect and simultaneously breaks down molecules, thereby greatly reducing background emission.

Another form of interelement effect encountered in emission spectrography is that one element can deactivate or deexcite another element, changing the emission intensity of the latter for a given concentration. This is particularly true in flame photometry but is not observed in plasma emission spectrography.

c. Matrix Effects

In atomic emission in general, matrix effects can be very important. For example, a sample of copper in a high boiling matrix will tend to vaporize preferentially, causing an efficient vaporization step. But if the copper is in a low boiling matrix, then the matrix preferentially vaporizes and the efficiency of vapor atomization of the copper is greatly reduced. Hence for the same concentration of copper in two matrixes, the emission intensity can be significantly different. This effect, known as the matrix effect, is always a problem in flame photometry and emission spectrography.

In the plasma emission the matrix effect is very much reduced. Usually the sample is introduced as a liquid. The components of the liquid are quickly vaporized and the sample atomized and excited in the very high temperature of the plasma both because of the high temperature and its long residence time. This greatly reduces the matrix effect in plasma emission spectroscopy.

D. INTERFACED ICP—MASS SPECTROMETER

1. Mass Spectroscopy

The conventional detector for the plasma emission spectrum is the photomultiplier. This detector is sensitive and exhibits fast response times. Unfortunately the emission spectrum of many elements is very complex, which leads to difficulties in identifying the lines to be measured for quantitative analysis and resolving lines close to each other to avoid spectral interference. Frequently spectral interference and background interference cannot be eliminated, and complicated, often error-prone sample pretreatments are necessary to correct for them.

The mass spectrometer has been developed and used as an alternate detector to the photomultiplier.

The mass spectrometer was used by J. J. Thompson in 1913 and Aston in 1919 to demonstrate that elements in the Periodic Table existed as various isotopes. This was a vital foundation stone in establishing the order of the elements in the Periodic Table. It explained why atomic weights were fractional (because elements consisted of mixtures of isotopes) rather than whole numbers. It also led to the confirmation of new elements discovered in the Periodic Table at the end of the nineteenth century. Since that time mass spectrometry has been devoted primarily to the analysis of organic materials.

First the large magnet mass spectrograph used by Aston was developed by Dempster to be more rapid and to cover a greater molecular range. More recently the work of McLafferty has led to improved resolution using double focusing. This led to the ability to generate empirical formulae formally based on molecular weights measured to seven or eight significant figures. In more recent years the multisector mass spec has further improved resolution in this type of instrument.

Parallel to these developments was the emergence of the time of flight mass spectrometer and the *quadropole mass spec*. The quadropole mass spec in particular has been highly exploited since it does not require that the entering ions have zero or near-zero lateral energy. This greatly facilitates the use of the quadropole mass spectrometer as a tandem instrument attached to some other system. It has been exploited particularly fruitfully in GC–mass spec combinations and more recently in HPLC–mass spec.

2. Advantages of Interfacing

The development of plasma mass spec using the quadropole has been a natural outcome of these developments. Some of the properties of major importance when combining these instruments include the fact that the ICP has a high ionization efficiency which approaches 100% for most of the elements in the Periodic Table but has a low incidence of doubly charged ions. The mass spectra are very simple and elements are easily identified from the atomic weights and the isotope ratios exhibited. This greatly relieves the major problem of the ICP, i.e., the rich complicated optical emission spectrum. In addition the mass spectrograph is very sensitive and linear over a wide dynamic range of up to five orders of magnitude. A list of compatible properties of ICP and MS was proposed by Hieftje et al. (12) as shown in Table 6.5. These attractive features have led to considerable attention being paid to interfacing these techniques.

Table 6.5 Complementary aspects of the ICP–MS

ICP emission spectrometry	Mass spectrometry
1. Efficient but mild ionization source (produces mainly singly charged ions).	Ion source required.
2. Sample introduction for solutions is rapid and convenient.	Sample introduction can be difficult for inorganic samples. Thermal, spark-source, or secondary ion sources are generally restricted to solid samples and are time-consuming.
3. Sample introduction is at atmospheric pressure.	Often requires reduced-pressure sample introduction.
4. Few matrix or interelement effects are observed and relatively large amounts of dissolved solids can be tolerated.	Limited to small quantities of sample.
5. Complicated spectra with frequent spectral overlaps.	Relatively simple spectra.
6. Detectability is limited by relatively high background continuum over much of the useful wavelength range.	Very low background level throughout a large section of the mass range.
7. Moderate sensitivity.	Excellent sensitivity.
8. Isotope ratios cannot usually be determined.	Isotope-ratio determinations are possible.

Source: Ref. 12.

The individual components of the ICP and mass spec have already been developed considerably. However, two major problems exist. The ICP operates at itostpheric pressure and at temperatures estimated to be in a range between 5,000 and 9,000 K. On the other hand, the mass spectrograph operates in a high vacuum u10⁻⁴ to 10⁻⁶ torr) and at room temperature. Interfacing of the two systems is therefore the critical problem to be overcome to provide the ideal marriage of these techniques.

3. Equipment

Simplistically the instrument is an ICP interfaced with a quadropole MS. The ICP and MS systems are standard. The interfacing is the difficult step.

a. The Process of Interfacing ICP Mass Spec

The earliest attempts to interface these systems utilized a simple metal interface (a skimmer) with a hole drilled through it with a diameter of approximately 60 μm. This was placed between the plasma and the MS. One side of the interface was exposed to the tip of the plasma and the other side to the entrance of the mass spectrometer. The interface was water cooled to prevent melting. It was anticipated that the elemental ions leaving the top of the plasma would enter the mass spectrometer, whereas much of the supporting argon gas would be discriminated against by the skimmer. Although this worked to some extent, unfortunately it became clear that it caused severe analytical problems. The water-cooled orifice created a layer of cool, slowly moving gas. Chemical reaction occurred in this layer, generating metal oxide, hydroxyl ions, and other more complex ions not present in the original plasma. Further, many of these ions were not elemental in composition and complicated the mass spectrum. Their generation depended on the plasma conditions utilized and the sample injected. The presence of dilute hydrochloric, nitric, or other acids in the sample caused further interferences to the mass spectrum.

Later interfaced systems utilized two orifices. The first was the *"sampler"* which was a water-cooled metal with an inverted cone shape. A circular hole with a diameter of between 0.5 and 1.0 mm was drilled through the center of the cone coaxially. A second metal cone, known as a *"skimmer,"* was placed immediately following the sampler, with a small gap between the two. The metal cone also contained a hole which led to the entrance of the mass spectrometer. This system is illustrated in Figure 6.12. Between the sampler and the skimmer was a region which was continuously pumped to about 1 torr. After passing through the skimmer the gases at this pressure enter the mass spectrometer at a reasonably low rate so that the mass spectrometer can be maintained at a low pressure.

Location of the system in the plasma was critical and has demanded close

To Mass Spectrometer

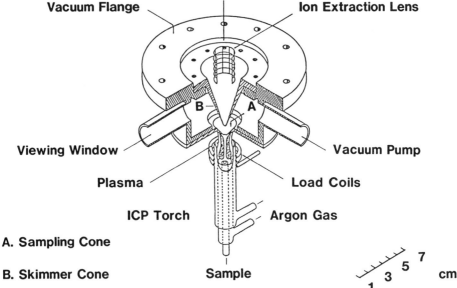

Vacuum Flange   Ion Extraction Lens

B   A

Viewing Window   Vacuum Pump

Plasma   Load Coils

ICP Torch   Argon Gas

A. Sampling Cone

7

B. Skimmer Cone   Sample   5   cm
                                                        3
                                                   1

Figure 6.12   ICP and ion-sampling interface. Note that the design adopted in this laboratory in conjunction with the Perkin-Elmer Corporation has a vertical configuration, whereas a horizontal configuration between the ICP and MS is more often used. Only the first ion extraction lens is shown. (From Ref. 12)

attention. At the present time the sampler is immersed in the analytical region of the plasma used for observation of the emission spectrum.

In early systems a secondary discharge occurred in the region between the two orifices. This led to the introduction of metal ions from the instrument into the mass spectrometer generating erroneous analytical data, also destruction of the instrument. In addition doubly charged ions were generated together with a photon background.

A series of ion-focusing elements similar to those developed in double-focusing mass spec have been utilized to introduce the ions into the quadropole. Also, photon blockers have eliminated interference effects due to the presence of large numbers of photons. Therefore, by suitable shielding and grounding these background signals have been largely eliminated in recent models.

4. Analytical Applications

One of the most important advantages of the ICP-MS is that it can be used to determine metals and nonmetals alike. Also, it can detect and measure positive and negative ions. As a consequence, nonmetals are measured as the negative ion and metals as the positive ion. It can be generalized that elements on the left-hand side of the Periodic Table are measured as positive ions and elements on the right-hand side are measured as negative ions. This capacity is a major advantage over both atomic absorption and all forms of emission spectrography. In practice, however, it has been found that only the halides can be detected with reasonable sensitivity (13) as shown in Tables 6.6 and 6.7. It was found that the background was higher than when positive ions were monitored, but the signal was about fifty times higher

The use of the quadropole mass spec has quickly borne fruit since it is known to be capable of rapidly scanning the large mass ranges encountered in organic analysis. Therefore, it was not difficult to scan the range up to 250 mass units in

Table 6.6 Comparison of the S/N Ratio for Selected Nonmetal Analytes in the Positive- and Negative-Ion Detection Modes

| Element | Isotope | Signal to background noise ratio[a] | |
		Positive ion	Negative ion
boron	11	1160	ND[b]
fluorine	19	ND	34.5
silicon	28	95.6	ND
phosphorus	31	507	ND
sulfur	34[c]	23.4	ND
chlorine	35	197	560
arsenic	75	8090	ND
selenium	78[c]	239	ND
bromine	79	843	856
tellurium	130	2300	ND
iodine	127	3770	431

[a]Analyte concentrations are 10 µg/mL.
[b]Not detectable.
[c]Isotopes that are not most abundant.
Based on Direct determination of nonmetals in solution by atomic spectroscopy, by D. A. Melpregor, K. B. Call, J. M. Gehlhausen, A. S. Viscomi, M. Wu, L. Zhang, and J. W. Carnahan. *Anal. Chem. 60* (19), 1089A (1988).

Table 6.7 Detection Limits for the Halogens in the Negative-Ion Mode

Element	Isotope	Detection limit[a] (ng/mL)
fluorine	19	400
chlorine	35	80
bromine	79	10
iodine	127	70

[a]S/N = 3, time constant = 1 s.
[b]F and Cl would have detection limits of 30 and 20 ng/mL, respectively, in a non-blank-limited situation.

Therefore, it was not difficult to scan the range up to 250 mass units in 20 milliseconds. It should be noted that 250 mass units covers most of the elements in the Periodic Table. Of course, the instruments can be used in a mode which jumps from one mass range to another without covering the entire range if so desired.

The detection limits reported so far indicate that in general the system is more sensitive than the ICP optical emission spectrograph and in many cases has a sensitivity comparable with that obtained in atomic absorption using a thermal atomizer. It is of course also better than the sensitivity obtained with atomic absorption obtained using flame atomizers. Detection limits in the order of 0.1–0.2 ppb are typical, whereas for ICP emission and flame atomic absorption typical results range between 1 and 40 ppb.

a. Dynamic Range

Using ion-counting measurement, a linear dynamic range of about five decades is achievable. This is much better than emission spectrography and atomic absorption spectroscopy and comparable to that obtained with ICP emission.

b. Isotope Measurements

A major new potential in the technique is the ability to measure the isotope ratios of the various elements. It is of interest to note that this was the original application of mass spectrometry both by Aston and by Thompson, and it is only now that we are again appreciating the analytical potential of this information. For example, the presence of isotopes and their relative abundance can be used to confirm the presence of the various elements.

Also, the application of ICP–mass spectroscopy to geological samples for the estimation of various isotopes has many potential applications, both for dating of rocks and for identification of original point of origin.

The system has been used for measuring the relative concentrations of various lead isotopes, and this can be used to identify the source of the lead since this varies from mine to mine (15). It has also been used for the determination of the following trace elements in food: Na, Mg, K, and Ca (16).

Work has also been carried out by Chong and R. S. Honk, who used argon ICP mass spec as a GC detector for nitrogen, oxygen, phosphorus, sulfur, carbon, chlorine, bromine, boron, and iodene.

c. Interferences

Mass spectral interferences are of course observed from argon ions since these are present in high abundance in plasma. In addition, ArH^+ and O^{2+} are present, and these may interfere with determinations of calcium and sulfur. The molecular ions from dilute solutions of reagent acids used in the original sample also cause problems. Ions such as ArN^+, ClO^+, $ArCl^+$, SO^+, and SO^{2t} are commonly observed. Based on studies of these interfering ions, nitric acid is considered to be the most attractive acid to be used in preparing sample solutions. No ions occur, however, with a mass greater than 82 from any of the common acids. Therefore, the higher part of the mass range is unaffected by these interfering ions.

Matrix Effects The introduction of the sample into the plasma suffers from the same problems in ICP mass spec as in ICP emission. These include the complicated process of nebulization and atomization. These processes occur before introduction into the plasma–mass spec system and are not eliminated by their position.

The interface between the two systems includes the 1 torr region where deposition of products can occur. Severe suppression of signal has been observed when high concentrations (100 µg per ml) are present. This may be caused by suppression of ionization by other elements present. At this time the cause of this interference is not clear, but the fact remains that interference does take place. The problem can be overcome to some extent by limiting the concentration in the samples to less than 0.2% total solids. This can be a serious limitation, particularly when body fluids are being examined.

Overlap of masses also occurs under varying conditions. For example, the isotopes MoO^+ and $MoOH^+$ ions overlap with cadmium (12).

A list of such interferences has been provided by M. A. Vaugh and G. Horlick.

An example of the spectra obtained from an interfaced plasma–MS system is shown in Figure 6.13. This illustrates a valuable procedure for the difficult task of rare earth analysis.

Detection limits for the method are shown in Table 6.8. In a recent development a system was produced using plasma–MS with an electrothermal sample introduc-

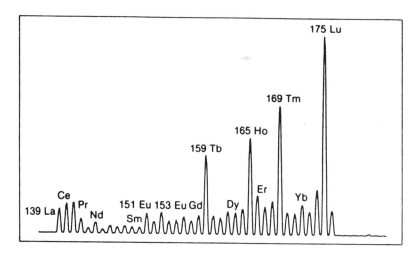

Figure 6.13 Spectra of a mixed solution of rare earths. Elemental analysis using plasma torch with mass spectrometer.

tion system. Detection limits of one to two orders of magnitude greater than normal are claimed, as illustrated in Table 6.9. This must be the ultimate in interfacing (17).

d. Limitations

The major limitations at this point seem to be the introduction system to the ICP (a common problem with the ICP), the matrix effects of solid materials dissolved in the sample, and mass ion overlap.

However, numerous applications have already been forthcoming such as to characterization of geochemical sample, alloys, steels, body fluids, foods and to the isotope ratio of the materials used in the nuclear industry. At present the only major manufacturer is Perkin-Elmer Corporation. It is hoped that the technique will continue to prosper.

E. CONCLUSIONS

There is no question but that the interfaced ICP–mass spec system holds many

Table 6.8 Relative Sensitivities Detection Limits (μg/L)

Element	Flame AA	Hg/Hydride	Furnace	ICP Emission	ICP–MS
Ag	0.9		0.005	1	0.04
Al	30		0.04	4	0.1
As	100	0.02	0.2	20	0.05
Au	6		0.1	4	0.1
B	700		20	2	0.1
Ba	8		0.1	0.1	0.02
Be	1		0.01	0.06	0.1
Bi	20	0.02	0.1	20	0.04
Br					1
C				50	50
Ca	1		0.05	0.08	5
Cd	0.5		0.003	1	0.02
Ce				10	0.01
Cl					1
Co	6		0.01	2	0.02
Cr	2		0.01	2	0.02
Cs	8		0.05		0.02
Cu	1		0.02	0.9	0.03
Dy	50				0.04
Er	40				0.02
Eu	20				0.02
F					100
Fe	3		0.02	1	0.2
Ga	50		0.1	10	0.08
Gd	1200				0.04
Ge	200		0.2	10	0.08
Hf	200				0.03
Hg	200	0.008	1	20	0.03
Ho	40				0.01
I					0.02
In	20		0.05	30	0.02
Ir	600		2	20	0.06
K	2		0.02	50	10
La	2000			1	0.01
Li	0.5		0.05	0.9	0.1
Lu	700				0.01
Mg	0.1		0.004	0.08	0.1
Mn	1		0.01	0.4	0.04
Mo	30		0.04	5	0.08
Na	0.2		0.05	4	0.06

(*Continued*)

Table 6.8   (*Continued*)

Element	Flame AA	Hg/Hydride	Furnace	ICP Emission	ICP–MS
Nb	1000			3	0.02
Nd	1000				0.02
Ni	4		0.1	4	0.03
Os	80				0.02
P	50000		30	30	20
Pb	10		0.05	20	0.02
Pd	20		0.25	1	0.06
Pr	5000				0.01
Pt	40		0.5	20	0.08
Rb	2		0.05		0.02
Re	500			20	0.06
Rh	4			20	0.02
Ru	70			4	0.05
S				50	500
Sb	30	0.1	0.2	60	0.02
Sc	20			0.2	0.08
Se	70	0.02	0.2	60	0.5
Si	60		0.4	3	10
Sm	2000				0.04
Sn	100		0.2	40	0.03
Sr	2		0.02	0.05	0.02
Ta	1000			20	0.02
Tb	600				0.01
Te	20	0.02	0.1	50	0.04
Th					0.02
Ti	50		1	0.5	0.06
Tl	9		0.1	40	0.02
Tm	10				0.01
U	10000			10	0.01
V	40		0.2	2	0.03
W	1000			20	0.06
Y	50			0.2	0.02
Yb	5				0.03
Zn	0.8		0.01	1	0.08
Zr	300			0.8	0.03

All detection limits were determined using elemental standards in dilute aqueous solution.

Atomic absorption (Model 5100) and ICP emission (Plasma II) detection limits are based on a 95% confidence level (2 standard deviations) using instrumental parameters optimized for the individual element. ICP emission detection limits obtained during multielement analyses will typically be within a factor of 2 of the values shown.

Cold vapor mercury AA detection limits were determined using a MHS-20 Mercury/Hydride system with an amalgamation accessory.

Furnace AA (Zeeman/5100) detection limits were determined using STPF conditions and are based on 100-μL sample volumes.

ICP-MS (ELAN 250) detection limits were determined on a multielement basis using a 98% confidence level (3 standard deviations). ICP-MS detection limits using operating conditions optimized for individual elements typically can be expected to be 3–10 times better than the values shown, depending on the element.

ICP-MS detection limits for fluorine and chlorine were determined using negative ion detection on the ELAN 250.

Source: Courtesy of Perkin-Elmer Corporation.

Table 6.9 Detection Limits

Elements	ETV-ICP-MS		graphite furnace AAS[a]		ETV-ICP-AES(2)	
	ng/mL (2 µL)	pg	ng/mL (10 µL)	pg	ng/mL (50 µL)	pg
^{107}Ag	0.08	0.16	0.01	0.1	0.1–30	1–300
^{27}Al	0.03	0.05	0.2	2	1.5–13000	7.5–26000
^{75}As	0.05	0.1	1	10	20–5000	200–10000
^{44}Ca	0.7	1.4	0.5	5	0.002–70000	0.02–139000
^{114}Cd	0.15	0.3	0.01	0.1	0.2–1640	1–3280
^{52}Cr	0.1	0.2	1	10	0.3–790	1.5–1580
^{39}K	1.5	3	0.1	1	110–83400	550–167000
^{23}Na	0.2	0.4	1	10	80–38000	400–76000
^{58}Ni	0.47	0.93	1	10	0.9–1050	4.5–2100
^{208}Pb	0.1	0.3	0.2	2	2–1400	20–2800
^{78}Se	5.7	11.4	1	10	6–600	600–3000
^{28}Si	2.7	5.4	5	50	10–500	100–2500
^{64}Zn	0.2	0.4	0.05	0.5	0.05–1700	0.25–3540

[a]JVL laboratory recent detection limits.

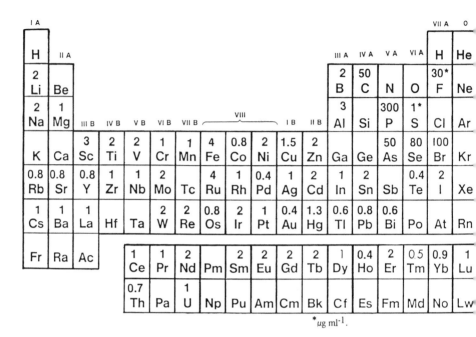

Figure 6.14   Detection limits using plasma MS analysis (ng/mL).

advantages over all existing techniques for elemantal analysis. There are problems to be solved at present, but these all seem to be within our capabilities. Typical sensitivities are shown in Figure 6.14.

A serious handicap with this system, however, is the high cost. This will prohibit routine use and general acceptance unless free competition is allowed to bring the price within the range of the pocketbook of most laboratories.

REFERENCES

1. Schrenk, W. G., *Applied Spec., 42, No. 1:* 4 (1988).
2. Northway, S. J., and R. C. Fry, *Applied Spec., 34(3):* 332 (1980).
3. Northway, S. J., and R. C. Fry, *Applied Spec., 34(3):* 338 (1980).
4. Marks, M. A., and G. N. Heiftje, *Applied Spec., 42:* 227 (1988).
5. Kumamanc, T., Y. Okamoto, and H. Matsuo, *Applied Spec., 47, No. 5:* 918 (1987).
6. Ceceonie, T., S. Mieraledharan, and H. Freiser, *Applied Spec., 42, No. 1:* 177 (1988).
7. Koropchak, J. A., and D. H. Winn, *Applied Spec., 41, No. 8:* 1311 (1987).
8. Belhorn, R. B., and M. B. Denton, *Applied Spec., 43.1:* 1 (1989).

9. Sheeline, A.,and D. N. Coleman, *Anal. Chem., 10:* 1185A (1987).

10. Pivoroku, D. E., W. G. Fateley, and R. C. Fry, *Applied Spec., 40(3):* 291 (1986).

11. Berthoud, T., P. Mauchien, A. Vian, and P. LeProvost, *Applied Spec., 44(5):* 913 (1988).

12. Selby, M., and G. M. Hiefje, *Am. Lab., 8:* 16 (1987).

13. Vickers, G. H., D. A. Wilson, and G. M. Heifje, *Anal. Chem., 60:* 1808 (1988).

14. Chong, N. S., and R. S. Honk, *Applied Spec., 41:* 66 (1987).

15. Muchlewicz, K. G. and J. W. Carnahan, *Anal. Chem., 58(14):* 3122 (1986).

16. Himmers, T. A., and E. Heithmar, *Anal. Chem., 59:* 00 (1987).

17. Satzger, R. D., *Anal. Chem., 60:* 2500 (1988).

18. Park, C. J., J. C. Van Loon, P. Arrowsworth, and J. B. French, *Anal Chem., 59:* 2191 (1987).

Index

A

Absorbance, 25
Absorption range, 121
  extension, 141
Accuracy, 37
Amplifier lock in, 85
Anion analysis, 285
Arc discharge, 223
Arsenic, 65
Ashing of sample, 232
Atomic absorption, 55
  compared to emission, 18, 25
  degree of, 90
  line width, 60, 61
  sensitivity, 61
Atomic fluorescence, 167
  math relationship, 168
Atomic spectroscopy, 1, 3, 55
Atomic theory, 5
  population, 20, 22
  size, 6
  structure, 6, 87
  weight, 6
Atomizers, 76
  carbon burner, 155
  carbon filament, 150
  electro thermal, 146
  flame, 77
  hollow T, 155
  L'vov platform, 158
  Lundegardt, 78, 80
  Massman, 151
  mechanical, 80, 95
  mini-Massman, 152
  tantalum boat, 158
  total consumption, 80
Atoms,
  free, 77, 91
  population, 91
Atomization, 92, 96, 196
  drop size, 97
  effect of flame temperature,
    98

[Atomization]
  effect of solvent, 98
  electro thermal, 146

B

Background absorption, 105
  correction, 107
  Smith-Heifje, 112
  Zeeman, 108, 173
Beer-Lambert Law, 24, 40
Blaze angle, 70
Bohr, Neils, 6, 7
Boltzmann distribution 11, 217
Bunsen and Kirchoff, 2
Burner,
  Lundegardh, 78, 80, 96, 196
  shielded, 198
  total consumption, 80, 196

C

Calibration curves, 142, 208, 277,
  114
  standard addition, 143
Carbon atomizer, 55
Chemical form of atoms, 98
Chemical interference, 102, 103,
  157, 212
Comparison,
  absorption and emission, 25, 259
Components of an atom, 5
Compton scattering, 12
Copper, 90
Counter electrodes, 226
Cross flow nebulizer, 258
Czerny-Turner mount, 253

D

Dalton, John, 5
Degree of absorption, 90
Detection limits, 117
  emission spectroscopy, 234
Detectors, 75
  fatigue, 86
  photomultiplier, 73
Direct line fluorescence, 176
Discharge,
  arc, 223
  spark, 225
Double monochromator, 253, 278
Drift, 29, 56
Drop Size (in atomizers), 198

E

Echelle monochromators, 222, 251
Electrodeless discharge lamp, 180
Electrodes, 226
Electromagnetic spectrum, 15
Electron configuration, 8, 9
  transitions, 11, 193
Electronic temperature, 11, 12, 217,
  243
Electrothermal atomizers, 146
Emission intensity, 22, 204
Emission spectroscopy, 215
Errors, 31, 38
Excitation interference, 212
Excited state, 87

F

Faraday, 215
Feussner, 216
Filter gas, 61, 63
Filter monochromators, 74, 200

[Filter monochromators]
  interference, 74
Flame absorption, 87
  energy, 196
  profile, 91
Flame photometry, 190
Flame temperature, 201
  ionization, 201
Flow injection, 82
Fluorescence, 167
  direct line, 176
  resonance, 175
  sensitized, 178
  thermally assisted, 178
Foucault, 1
Fraunhofer, 1, 215
Free atoms, 17

G

Gratings, 69, 222
  blazing, 70
  orders, 70
  order sorter, 71
Ground state, 87
Grotrian diagram, 88, 194

H

Halides, 17, 89
Hartley, 216
Hollow cathodes, 61
  demountables, 64
  hardening, 64
  intensity, 62
  multielement, 66
Holograph, 74

I

Interfacing MS, 282
Interference filters, 74
Interferences, 211, 278
  background, 105, 237, 278
  chemical, 212
  excitation, 212
  radiation, 211
Internal standards, 236
Ionization in flames, 201
Isotope analysis, 296

L

Laser excitation, 173, 180
Lead, 63
Lecog, 216
Leeman ICP, 263
Line reversal, 207
Littrow prism, 220
Lock in amplifier, 85
Lockyer, 215
L'vov, 56
  platform, 158

M

Margoshes, M., 241
Martin, A. J. P., 2
Massman,
  atomizer, 151
  mini-Massman, 152
Mass spectrometry, 281
  interfacing, 282
Mechanical chopper, 85
Mercury, 65
Methanol burner, 95
Mirrors, 199

Modulation, 83, 84, 181
Monochromator, 66, 199, 218
Multielement analysis, 183

N

Nebulizer, 258
Nitrous oxide flame, 192, 201
noise, 47, 49, 75

O

Operation error, 4
Optical system, 26, 56
    double beam, 21, 58
    single beam, 26, 56
Oscillator strength, 90

P

Phase (light), 13
Photoelectron effect, 12
Photomultiplier, 75
Planck, 13
Plasma emission, 241
    compared to,
    DC plasma, 241, 250
    emission spectroscopy, 245
    ICP, 241
Plasma torch, 246
Polychromator, 246
Preburn time, 236
Precision, 37
Prism, 68, 219

Q

Qualitative analysis, 230
Quantitative analysis, 140, 232,
    264, 277

R

Radiation, 12
    analytical field, 16
    source, 60
Radiation interference, 211
Raies ultima, 231, 240
Refractal plate, 256
Reliability of results, 30
Resolution, 68
    grating, 69
    prism, 69
Resonance line, 88
    fluorescence, 175
Response curves, 233, 245
Ringbom plot, 40
Rotating disc electrode, 228
Rowland Circle, 222, 250

S

Sampling, 50
Sauter mean, 198
Selenium, 65
Self absorption, 62
Semi-quantitative analysis, 238
Sensitivity, 117
    atomic absorption, 185
    atomic fluorescence, 185
    flame photometry, 190, 192
Separated flames, 172
Sequential analysis, 232
Signal,
    integration, 236

[Signal]
-to-noise ratio, 47, 49, 75, 76
Significant figures, 45
Skimmer ICP-MS, 283
Slit, 67, 115
  mechanical, 67
  spectral, 67
  smoothing condensers, 229
Solvent effect, 92, 202
Spark discharge, 225,
Spattering, 61, 64
Standard addition, 143
Standard deviation, 36
Stark effect, 60
Stationary states, 6
Statistical definition, 35
Step section, 239
Steppers motor, 256
Stolov, A. L., 243
Surface tension, 198

T

Temperature control (ICP), 264
Thermospray, 83, 198, 258
Transmittance, 24

U

Ultrasonic nebulizer, 83

V

Viscometry, 198

W

Walsh, Alan, 2, 61, 84
Wobbler, 201
Wollaston, 1

Z

Zeeman background correction, 108, 173
Zeeman effect, 60